The
FOURTH PHASE
OF WATER

BEYOND
SOLID, LIQUID, AND VAPOR

THE FOURTH PHASE OF WATER

BEYOND SOLID, LIQUID, AND VAPOR

GERALD H. POLLACK

EBNER & SONS PUBLISHERS
EBNERANDSONS.COM
SEATTLE WA, USA

Ebner and Sons Publishers
3714 48th Avenue NE
Seattle, WA 98105
www.ebnerandsons.com
info@ebnerandsons.com

For quantity orders, please contact publisher.

Library of Congress Control Number: 2012956209

Pollack, Gerald H., PhD

The Fourth Phase of Water: Beyond Solid, Liquid, and Vapor

ISBN Hardcover: 978 – 0 – 9626895 – 3 – 6
ISBN Paperback: 978 – 0 – 9626895 – 4 – 3 (this edition)
ISBN ePUB ebook: 978 – 0 – 9626895 – 5 – 0
ISBN Kindle ebook: 978 – 0 – 9626895 – 6 – 7
ISBN PDF ebook: 978 – 0 – 9626895 – 7 – 4

9 8 7 6 5 4 3 2

Printed in South Korea
Fonts: Birmingham, IowanOldSt BT, Calibri

Cover design by Ethan Pollack

Layout design by Amanda Fredericks and Ethan Pollack

Illustrated by\Ethan Pollack

to Gilbert Ling

who taught me that water in the cell
is nothing like water in a glass;

whose courage has been a
continuing inspiration.

Acknowledgments

Like a child raised by a village, the production of this book evolved from the efforts of numerous loosely connected people.

Foremost must certainly be Gilbert Ling, whose monumental contributions first stirred my interest in water. Ling has been far ahead of his time. His pioneering work opened the eyes of many scientists to the realization that water is not merely a background carrier of the common molecules of life; it is a central player in all of life's processes. Sadly, his many contributions have gone unrecognized, and his willingness to challenge science at its core has made him something of a pariah. From the time I first met Gilbert in the mid-1980s, he has continued to inspire me; if anyone is responsible for seeding the creation of this book, that person is Gilbert Ling.

Second in line: Vladimir Voeikov, of Moscow University. Few scientific subjects exist that Vladimir does not know a great deal about, and I must confess that many of the issues considered in this book began from conversations with him. His wide-ranging insights have broadened my peripheral vision. I thank him also for cooking up a series of marvelous Russian dinners during the time we were ensconced in a St. Petersburg apartment. The combination of pilmeni and vodka stirred the creative neurons to fire so rapidly that they could probably be sensed as far as Chicago.

As for the actual preparation of this book, I'm indebted mainly to three people, listed in chronological order of their contributions.

The first, Brandon Reines, helped even before we'd met. Brandon and I had enjoyed a long-standing scientific correspondence. When he raised a daunting question, I suggested that reading a preliminary draft of my forthcoming book might provide the answer. Brandon took the plunge. He responded that the book seemed too much of a good thing: one hot-fudge sundae for dessert can be fine, but fifteen of them are too much for anyone to digest. Brandon's "culinary" input spares you chapters on subjects ranging from how birds fly (not as you think) all the way to why the life spans among species can range from a few days to a few thousand years. I will reserve those and various other subjects for subsequent books. Brandon has been spectacularly helpful in packaging each chapter's material in reader-friendly ways, in composing headers that don't provoke yawns, and in countless additional ways that have made a real difference. For all of that, I (and my readers) owe him a huge debt of gratitude.

Second is the artist — my son Ethan Pollack. Already sketching at the age of four, Ethan went on to study sculpture at Syracuse University, developed his skills during a stint in Florence, apprenticed in New York with the world-class artist Jeff Koons — and finally returned home to Seattle. Working with him has been a pure delight. Ethan has shown a deep understanding of the relevant scientific concepts, a high level of sensitivity, an unusual degree of creativity, a strong inclination to pay attention to detail, and an unrelenting dedication to the project's overall success. If you find the concepts clearly and attractively illustrated, Ethan is the person to thank.

Finally, I thank my editor, Don Scott. Don is one of the most articulate persons I know. A philosopher by education and an attorney by training, Don has a special knack for words. Consistently, he could divine what I was struggling to say but couldn't quite figure out how. He suggested nuanced phrases for the clumsy ones I'd drafted. And he showed an uncanny ability to spot gaps in logic, even in material lying well beyond his sphere of expertise. If some obscurities remain, it is probably because of my stubborn inclination to ignore his advice.

Beyond those major contributors, I had the benefit of three cohorts of reviewers, one of which was my laboratory's staff members. Lab members were not shy about telling me which aspects of the book they did not agree with. Several expressed discomfort with the reach of some of my unorthodox proposals; hence, I made it clear that the responsibility lies completely with me, not with them. Their incisive commentary, conveyed during a series of lunchtime feedback sessions and text annotations, helped reshape the final draft — especially some of the more difficult chapters. It goes without saying that their many experimental contributions form the skeletal framework of this entire book.

Undergraduates provided similarly useful feedback. The laboratory attracts a sizeable group of volunteer undergraduate researchers. To many of them, experiments seem more like play than work: we supply the toys, and they use their imagination to pursue experiments that scientific "adults" might not dare contemplate. Undergraduates love those experiments. Some of them turned up results completely unexpected, a few even seminal. This book presents those findings in some detail. Besides their experimental contributions, many of those undergraduates read and critiqued successive versions of the text, for which I offer my sincere thanks.

Beyond these two sets of reviewers, many colleagues worldwide critiqued early drafts of the manuscript. Those colleagues ranged from chemists, physicists, and engineers all the way to biological scientists and even a handful of nonscientists. Some spent many hours. Their collective advice has prevented me from wandering too far astray. It also helped me to organize the material — a task less simple than you might guess. Several noted that the goal of producing a single volume that accounts for the totality of water science was nigh unto impossible: each chapter could expand into an entire book, and deciding on an appropriate balance between readability and length was a challenging task.

For altogether different reasons, I thank my family members. To my life-partner Emily Freedman, a public admission: I broke the promise made in my previous book — that the next book would be shorter and would consume less of my time. This book is longer; and I have two additional books in the works. Emi has been exceptionally understanding of the consuming demands required by projects of this magnitude. She has shown the patience of an angel. The rest of my family has been supportive as well: Mia, having to practically make an appointment to distract her father from his mesmerizing laptop; Ethan, graciously accommodating my many demands for artwork revision with a consistently upbeat attitude; and Seth, forever humoring me about my inclination to answer any question with: "structured water." My family could not have been more supportive during this multiple-year endeavor.

Finally, from the above-mentioned cohorts, I want to list all those who have critiqued the manuscript or portions thereof. This list includes students, research fellows, scientists, and some lay people. Their level of helpfulness was frequently out of proportion to their academic status; hence, I will list these contributors alphabetically, and if someone's name has been inadvertently omitted I surely apologize.

I offer my thanks to the following people: Peter Allen, Brandon Bowman, Brian Biccum, Frank Borg, Binghua Chai, Ruying Chen, Daniel Chiang, Chi Chuang, Cara Comfort, Charles Cushing, Ronnie Das, Ken Davidson, James deMeo, Aparajeeta Duttchoudhury, Nigel Dyer, Collin Eddington, Xavier Figueroa, Herb Fleschner, Ben Flowers, Emily Freedman, Gonzalo Garcia, Karl Gatterer, Matthew Gelber, Krystal Ginter, Matias Gonzalez, Ron Griffin, John Grigg, Zhanna Grigoryan, Emmanuel Haven, Maewan Ho, Arie Horowitz, Linda Hufnagel, Breanna Huschka, John Hwang, Federico Ienna, Hiromasa Ishiwatari, Tengiz Jaliashvili, Manal Jmaileh, Konstantin Korotkov, Ethan Kung, Kurt Kung, Victor Kuz, Alysia Letourneau, Zheng Li, Molly McGee, Lior Miller, Francesco Musumeci, Kylie van Nguyen, Derek Nhan, Gabriela Patilea, Bernard Pennock, Ari Penttila, Orion Polinsky, Ethan Pollack, Seth Pollack, Sylvia Pollack, Leo Ramakers, Randy Randall, Sudeshna Sawoo, Rainer Stahlberg, Clint Stevenson, Heather Swain, Masaaki Takarada, Shrutee Tandon, Yolene Thomas, Tony Thomson, Merry Toh, Gerard Trimberger, Karoly Trombitas, Outi Villet, Vladimir Voeikov, Jacob Woller, Jeff Yang, Hyok Yoo, and Rolf Ypma. Three of these — Cara Comfort, Charles Cushing, and Rolf Ypma — put in an exceptional amount of time and effort.

Finally, I thank Amanda Fredericks for her creative input into layout and her attention to detail, and Rolf Ypma for his scrupulous work on indexing.

Producing this book took the combined efforts of a large community of caring and thoughtful people. I warmly thank all of those who have contributed.

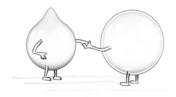

Contents

Preface

There in my living room sat the Nobel laureate. He was shy and I was intimidated, a combination certain to generate awkwardness. It was like trying to make small talk with Einstein. What do you say?

Sir Andrew Huxley was a Nobelist among Nobelists. He had already done classical work on cell membranes, and by the time of our meeting he'd become the leader in the field of muscle contraction. His many accolades included President of the Royal Society; Master of Trinity College, Cambridge; and recipient of the Order of Merit from the Queen of England. He was also a member of the distinguished Huxley family, a lineage that produced the legendary biologist Thomas Henry Huxley ("Darwin's bulldog") and the prescient writer Aldous Huxley. Here in my humble living room sat this towering scientific aristocrat.

During those awkward moments, nobody dared mention the elephant in the room: experimental results from our laboratory demonstrating that my guest's theory might be wrong. He'd come to check out our evidence, which took place earlier, within the confines of my laboratory. But, in my living room, we avoided that thorny subject altogether, focusing instead on such compelling issues as the weather. Even with a few rounds of sherry for social lubrication, it was a struggle to let it hang out; after all, Huxley was a scientific oracle — practically a deity.

Towering figures like Huxley appear awesome; however, we tend to forget that even the most renowned scientists are human. They eat the same foods we eat, share the same passions, and are subject to the same human foibles. So, while we may marvel at their insights and respect their contributions, we need not feel obliged to treat those contributions as faultless or absolute; scientific formulations are hardly sacred.

Treating any scientific formulation as sacred is a serious error. Any framework of understanding that we build needs to rest on solid foundations of experimental evidence rather than on sacred formulations; otherwise, the finished product may resemble one of M.C. Escher's renderings of subtle impossibility — a result worth avoiding. Even long-standing models remain vulnerable if they have not managed to bring simple, satisfying understandings. Galileo's story teaches us that when an established foundation requires the support of elaborate "epicycles" to agree with empirical observations, it's time to begin searching for simpler foundations.

This book attempts to build reliable foundations for a new science of water. The foundation derives from recent discoveries. Upon this new foundation, we will build a framework of understanding with considerable predictive power: everyday phenomena become plainly explainable without the need for mind-bending twists and jumps. Then comes the bonus: the process of building this new framework will yield four new scientific principles — principles that may prove applicable beyond water and throughout all of nature.

Thus, the approach I take is unconventional. It does not build on the "prevailing wisdom"; nor does it reflexively accept all current foundational principles as inherently valid. Instead, it returns to the root method of doing science — relying on common observation, simple logic, and the most elementary principles of chemistry and physics to build understanding. Example: in observing the vapor rising from your cup of hot coffee, you can actually *see* the clouds of vapor. What must that tell you about the nature of the evaporative process? Do prevailing foundational principles sufficiently explain what you see? Or must we begin looking elsewhere? (You'll know what I mean if you read Chapter 15.)

This old-fashioned approach may come across as mildly irreverent because it pays little homage to the "gods" of science. On the other hand, I believe the approach may provide the best route toward an intuitive understanding of nature — an understanding that even laymen can appreciate.

I certainly did not begin my life as a revolutionary. In fact, I was pretty conventional. As an undergraduate electrical engineering student, I came to class properly dressed and duly respectful. At parties, I wore a tie and jacket just like my peers. We looked about as revolutionary as members of an old ladies' sewing circle.

Only in graduate school at the University of Pennsylvania did someone implant in me the seeds of revolution. My field of study at the time was bioengineering. I found the engineering component rather staid, whereas the biological component brought some welcome measure of leavening. Biology seemed the happening place; it was full of dynamism and promise for the future. Nevertheless, none of my biology professors even hinted that students like us might one day create scientific breakthroughs. Our job was to add flesh to existing skeletal frameworks.

I thought that incrementally adding bits of flesh was the way of science until a colleague turned on the flashing red lights. Tatsuo Iwazumi arrived at Penn when I was close to finishing my PhD. I had built a primitive computer simulation of cardiac contraction based on the Huxley model, and Iwazumi was to follow in my footsteps. "Impossible!" he asserted. Lacking the deferential demeanor characteristic of most Japanese I'd known, Iwazumi stated in no uncertain terms that my simulation was worthless: it rested on the accepted theory of muscle contraction, and that theoretical mechanism couldn't possibly work. "The mechanism is intrinsically unstable," he continued. "If muscle really worked that way, then it would fly apart during its very first contraction."

Whoa! A frontal challenge to Huxley's muscle theory? No way.

Although (the late) Iwazumi exuded brilliance at every turn and came with impeccable educational credentials from the University of Tokyo and MIT, he seemed no match for the legendary Sir Andrew Huxley. How could such a distinguished Nobel laureate have so seriously erred? We understood that the scientific mechanisms announced by such sages constituted ground truth and textbook fact, yet here came this brash young Japanese engineering student telling me that this particular truth was not just wrong, but impossible.

Reluctantly, I had to admit that Iwazumi's argument was persuasive — clear, logical, and simple. As far as I know, it stands unchallenged to this very day. Those who hear the argument for the first time quickly see the logic, and most are flabbergasted by its simplicity.

For me, this marked a turning point. It taught me that sound logical arguments could trump even long-standing belief systems buttressed by armies of followers. Once disproved, a theory was done — finished.

The belief system was gone forever. Clinging endlessly was tantamount to religious adherence, not science. The Iwazumi encounter also taught me that thinking independently was more than just a cliché; it was a necessary ingredient in the search for truth. In fact, this very ingredient led to my muscle-contraction dispute with Sir Andrew Huxley (which never did resolve).

Challenging convention is not a bed of roses, I assure you. You might think that members of the scientific establishment would warmly embrace fresh approaches that throw new light on old thinking, but mostly they do not. Fresh approaches challenge the prevailing wisdom. Scientists carrying the flag are apt to react defensively, for any such challenge threatens their standing. Consequently, the challenger's path can be treacherous — replete with dangerous turns and littered with formidable obstacles.

Obstacles notwithstanding, I did somehow manage to survive during those early years. By delicately balancing irreverence with solid conventional science and even a measure of obeisance, I could press on largely unscathed. Our challenges were plainly evident, but we pioneered techniques impressive enough that my students could land good jobs worldwide, some rising to academia's highest levels. Earning that badge of respectability saved me from the terminal fate common to most challengers.

During the middle of my career, my interests began expanding. I sniffed more broadly around the array of scientific domains, and as I did I began smelling rats all over. Contradictions abounded. Some of the challenges I saw others raise to their fields' prevailing wisdom seemed just as profound as the ones raised in the muscle-contraction field.

One of those challenges centered on the field of water — the subject of this book. The challenger of highest prominence at the time was Gilbert Ling. Ling had invented the glass microelectrode, which revolutionized cellular electrophysiology. That contribution should have earned him a Nobel Prize, but Ling got into trouble because his results began telling him that water molecules inside the cell lined up in an orderly fashion. Such orderliness was anathema to most biological and physical scientists. Ling was not shy about broadcasting his conclusions, especially to those who might have thought otherwise.

So, for that and other loudly trumpeted heresies, Ling eventually fell from favor. Scientists holding more traditional views reviled him as a provocateur. I thought otherwise. I found his views on cell water to be just as sound as Iwazumi's views on muscle contraction. Unresolved issues remained, but on the whole his proposal seemed evidence-based, logical, and potentially far-reaching in its scope. I recall inviting Ling to present a lecture at my university. A senior colleague admonished me to reconsider. In an ostensibly fatherly way, he warned that my sponsorship of so controversial a figure could irrevocably compromise my own reputation. I took the risk — but the implications of his warning lingered.

Ling's case opened my eyes wider. I began to understand why challengers suffered the fates they did: always, the challenges provoked discomfort among the orthodox believers. That stirred trouble for the challengers. I also came to realize that challenges were common, more so than generally appreciated. Not only were the water and muscle fields under siege, but voices of dissent could also be heard in fields ranging from nerve transmission to cosmic gravitation. The more I looked, the more I found. I don't mean flaky challenges coming from attention-seeking wackos; I'm referring to the meaningful challenges coming from thoughtful, professional scientists.

Serious challenges abound throughout science. You may be unaware of these challenges, just as I had been until fairly recently, because the challenges are often kept beneath the radar. The respective establishments see little gain in exposing the chinks in their armor, so the challenges are not broadcast. Even young scientists entering their various fields may not know that their particular field's orthodoxy is under siege.

The challenges follow a predictable pattern. Troubled by a theory's mounting complexity and its discord with observation, a scientist will stand up and announce a problem; often that announcement will come with a replacement theory. The establishment typically responds by ignoring the challenge. This dooms most challenges to rot in the basement of obscurity. Those few challenges that do gain a following are often dealt with aggressively: the establishment dismisses the challenger with scorn and disdain, often charging the poor soul with multiple counts of lunacy.

The consequence is predictable: science maintains the *status quo*. Not much happens. Cancer is not cured. The edifices of science continue to grow on weathered and sometimes even crumbling foundations, leading to cumbersome models and ever-fatter textbooks filled with myriad, sometimes inconsequential details. Some fields have grown so complex as to become practically incomprehensible. Often, we cannot relate. Many scientists maintain that that's just the way modern science must be — complicated, remote, separated from human experience. To them, cause-and-effect simplicity is a quaint feature of the past, tossed out in favor of the complex statistical correlations of modernity.

I learned a good deal more about our acquiescence to scientific complexity by looking into Richard Feynman's book on quantum electrodynamics, aptly titled *QED*. Many consider Feynman, a legendary figure in physics, the Einstein of the late 20th century. In the Introduction to the 2006 edition of Feynman's book, a prominent physicist states that you'll probably not understand the material, but you should read the book anyway because it's important. I found this sentiment mildly off-putting. However, it was hardly as off-putting as what Feynman himself goes on to state in his own Introduction: "It is my task to convince you *not* to turn away because you don't understand it. You see, my physics students don't understand it either. That's because *I* don't understand it. Nobody does."

The book you hold takes an approach that challenges the notion that modern science must lie beyond human comprehension. We strive for simplicity. If the currently accepted orthodox principles of science cannot readily explain everyday observations, then I am prepared to declare that the emperor has no clothes: these principles might be inadequate. While those foundational principles may have come from towering scientific giants, we cannot discount the possibility that new foundations might work better.

Our specific goal is to understand water. Water now *seems* complicated. The understanding of everyday phenomena often requires complex twists and non-intuitive turns — and still we fail to reach satisfying understandings. A possible cause of this unsatisfying complexity is the present foundational underpinning: an *ad hoc* collection of long-standing principles drawn from diverse fields. Perhaps a more suitable foundation — built directly from studying water — might yield simpler understandings. That's the direction we're headed.

To read this book, you needn't be a scientist; the book is designed for anyone with even the most primitive knowledge of science. If you understand that positive attracts negative and have heard of the periodic table, then you should be able to get the message. On the other hand, those who might thumb their noses at anything that seriously questions current dogma will certainly find the approach distasteful, for threads of challenge weave through the book's very fabric. This book is unconventional —a saga filled with steamy scenes and unexpected twists, all of which resolve into something I hope you will find satisfying, and perhaps even fun to read.

I have restricted formal references to those instances in which citations seemed absolutely necessary. Where the point is generally known or easily accessible, I've omitted them. The overarching goal was to streamline the text for readability.

Finally, let me admit to having no delusion that all of the ideas offered here will necessarily turn out to be ground truth. Some are speculative. I have certainly aimed at producing science fact, not science fiction. However, as you know, even a single ugly fact can demolish the most beautiful of theories. The material in this book represents my best and most earnest attempt to assemble the available evidence into a cohesive interpretational framework. The framework is unconventional, and I already know that some scientists do not agree with all aspects. Nevertheless, it is a sincere attempt to create understanding where little exists.

So, as we plunge into these murky waters, let us see if we can achieve some needed clarity.

GHP

Seattle, September 2012

Discovery consists of seeing what everybody has seen and thinking what nobody has thought.

Albert Szent-Györgyi,
Nobel laureate (1893-1986)

A BESTIARY

A READER'S GUIDE TO THE SPECIES THAT LURK
WITHIN THE MYSTERIOUS AQUEOUS DOMAIN

WATER MOLECULE

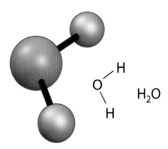

H₂O

The familiar water molecule, composed of two hydrogen atoms and one oxygen atom.

BULK WATER

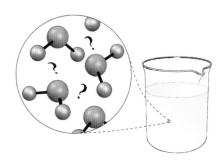

The standard collection of water molecules, whose arrangement is still debated.

EXCLUSION ZONE (EZ)

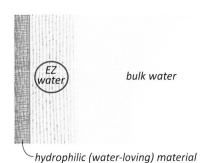

EZ water

bulk water

hydrophilic (water-loving) material

The "exclusion zone" (EZ), the unexpectedly large zone of water that forms next to many submersed materials, got its name because it excludes practically everything. The EZ contains a lot of charge, and its character differs from that of bulk water. Sometimes it is referred to as water's fourth phase.

ELECTRON AND PROTON

Electrons and protons are the elementary units of charge. They attract one another because one is positive and the other is negative. Electrons and protons play central roles in water's behavior — more than you might think.

WATER MOLECULE CHARGE

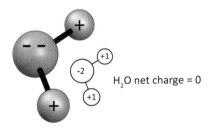

H_2O net charge = 0

The water molecule is neutral. Oxygen has a charge of minus two, while each of the hydrogen atoms has a plus one charge.

HYDRONIUM ION

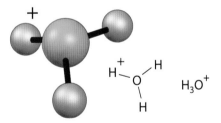

H_3O^+

Protons latch onto water molecules to form hydronium ions. Imagine a positively charged water molecule and you've got a hydronium ion. Charged species like hydronium ions are highly mobile and can wreak much havoc.

Interfacial Battery

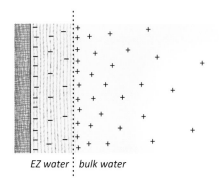

EZ water : bulk water

This battery comprises the exclusion zone and the bulk water zone beyond. The respective zones are oppositely charged, and the separation is sustained, as in an ordinary battery.

Radiant Energy

Radiant energy charges the battery. The energy comes from the sun and other radiant sources. The water absorbs these energies and uses them to charge the battery.

Honeycomb Sheet

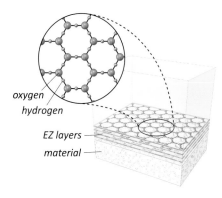

oxygen
hydrogen
EZ layers
material

The honeycomb sheet is the EZ's unitary structure. Sheets stack parallel to the material surface to build the EZ.

Ice

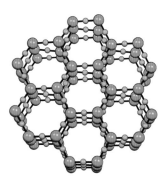

The atomic structure of ice closely resembles the atomic structure of the exclusion zone. This similarity is beyond coincidence: one transforms readily into the other.

DROPLET

The water droplet consists of an EZ shell that envelops bulk water. The two components have opposite charges.

BUBBLE

The bubble is structured like the droplet, except that it has a gaseous interior. Commonly, that gas is water vapor.

VESICLE

Since droplets and bubbles are similarly constructed, we introduce the generic label: *vesicle*. A vesicle can be a droplet or a bubble, depending on the phase of the water inside. When a droplet absorbs enough energy, it can become a bubble.

SECTION I

Water Riddles:
Forging the Pathway

1 Surrounded by Mysteries

Beaker in hand, two students rushed down the hall to show me something unexpected. Unfortunately, their result vanished before I could take a look. But it was no fluke. The next day the phenomenon reappeared, and it became clear why the students had reacted with such excitement: they had witnessed a water-based phenomenon that defied explanation.

Water covers much of the earth. It pervades the skies. It fills your cells — to a greater extent than you might be aware. Your cells are two-thirds water by volume; however, the water molecule is so small that if you were to count every single molecule in your body, 99% of them would be water molecules. That many water molecules are needed to make up the two-thirds volume. Your feet tote around a huge sack of mostly water molecules.

What do we know about those water molecules? Scientists study them, but rarely do they concern themselves with the large ensembles of water molecules that one finds in beakers. Rather, most scientists focus on the single molecule and its immediate neighbors, hoping to extrapolate what they learn to larger-scale phenomena that we can see. Everyone seeks to understand the observable behavior of water, i.e., how its molecules act "socially."

Do we really understand water's social behavior?

Since water is everywhere, you might reasonably conclude that we understand it completely. I challenge you to confirm that common presumption. Below, I present a collection of everyday observations, along with a handful of simple laboratory observations. See if you can explain them. If you can, then I lose; you may stop reading this book. If the explanations remain elusive even after consulting the abundant available sources, then I ask you to reconsider the presumption that we know everything there is to know about water.

I think we don't. Let's see how you fare.

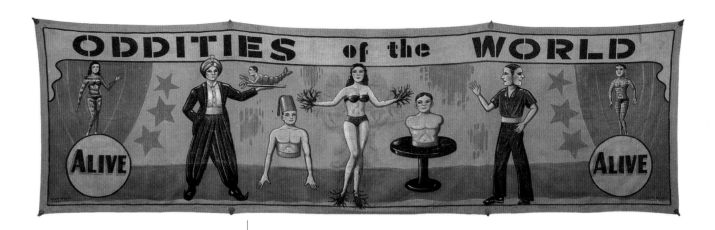

Everyday Mysteries

Here are fifteen everyday observations. Can you explain them?

• *Wet sand vs. dry sand.* When stepping into dry sand, you sink deeply, but you hardly sink into the wet sand near the water's edge. Wet sand is so firm that you can use it for building sturdy castles or large sand sculptures. The water evidently serves as an adhesive. But how exactly does water glue those sand particles together? (The answer is revealed in Chapter 8.)

• *Ocean waves.* Waves ordinarily dissipate after traveling a relatively short distance. However, tsunami waves can circumnavigate the Earth several times before finally petering out. Why do they persist for such immense distances? (See Chapter 16.)

• *Gelatin desserts.* Gelatin desserts are mostly water. With all that water inside, you'd expect a lot of leakage (**Fig. 1.1**). However, none occurs. Even from gels that are as much as 99.95% water,[1] we see no dribbling. Why doesn't all that water leak out? (Read Chapters 4 and 11.)

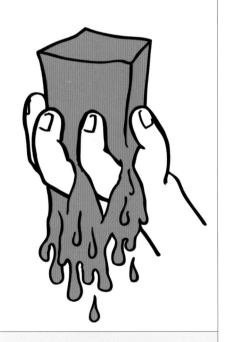

Fig. 1.1 *What keeps the water from dribbling out of the Jell-O?*

• *Diapers.* Similar to gels, diapers can hold lots of water: more than 50 times their weight of urine and 800 times their weight of pure water. How can they hold so much water? (Look at Chapter 11.)

• *Slipperiness of ice.* Solid materials don't usually slide past one another so easily: think of your shoes planted on a hilly street. Friction keeps you from sliding. If the hill is icy, however, then you must exercise great care to keep from falling on your face. Why does ice behave so differently from most solids? (Chapter 12 explains.)

- *Swelling.* Your friend breaks her ankle during a tennis match. Her ankle swells to twice its normal size within a couple of minutes. Why does water rush so quickly into the wound? (Chapter 11 offers an answer.)

- *Freezing warm water.* A precocious middle-school student once observed something odd in his cooking class. From a powdered ice cream mix he could produce his frozen treat faster by adding warm water instead of cold water. This paradoxical observation has become famous. How is it that warm water can freeze more rapidly than cold water? (See Chapter 17.)

- *Rising water.* Leaves are thirsty. In order to replace the water lost through evaporation in plants and trees, water flows upward from the roots through narrow columns. The commonly offered explanation asserts that the tops of the columns exert an upward drawing force on the water suspended beneath. In 100-meter-tall redwood trees, however, this is problematic: the weight of the water amassed in each capillary would suffice to break the column. Once broken, a column can no longer draw water from the roots. How does nature avert this debacle? (Check out Chapter 15.)

- *Breaking concrete.* Concrete sidewalks can be cracked open by up-welling tree roots. The roots consist mainly of water. How is it possible that water-containing roots can exert enough pressure to break slabs of concrete? (Look through Chapter 12.)

- *Droplets on surfaces.* Water droplets bead up on some surfaces and spread out on others. The degree of spread serves, in fact, as a basis for classifying diverse surfaces. Assigning a classification, however, doesn't explain *why* the droplets spread, or *how far* they spread. What forces cause a water droplet to spread? (Go to Chapter 14.)

- *Walking on water.* Perhaps you've seen videos of "Jesus Christ" lizards walking on pond surfaces. The lizards scamper from one end to the other. Water's high surface tension comes to mind as a plausible

Fig. 1.2 *What directs the rising water vapor to specific locations?*

explanation, but if surface tension derives from the top few molecular layers only, then that tension should be feeble. What is it about the water (or about the lizard) that makes possible this seemingly biblical feat? (Read Chapter 16.)

• *Isolated clouds.* Water vapor rises from vast uninterrupted reaches of the ocean's water. That vapor should be everywhere. Yet puffy white clouds will often form as discrete entities, punctuating an otherwise clear blue sky (**Fig. 1.2**). What force directs the diffuse rising vapor towards those specific sites? (Chapters 8 and 15 consider this issue.)

• *Squeaky joints.* Deep knee bends don't generally elicit squeaks. That's because water provides excellent lubrication between bones (actually, between cartilage layers that line the bones). What feature of water creates that vanishingly small friction? (Take a look at Chapter 12.)

• *Ice floats.* Most substances contract when cooled. Water contracts as well — until 4 °C. Below that critical temperature water begins expanding, and very much so as it transitions to ice. That's why ice floats. What's special about 4 °C; and, why is ice so much less dense than water? (Chapter 17 answers these questions.)

• *Yoghurt's consistency.* Why does yoghurt hold together as firmly as it does? (See Chapter 8.)

Mysteries from the Laboratory

I next consider some simple laboratory observations, beginning with the one seen by those students rushing down the hall to show me what they'd found.

(i) The Mystery of the Migrating Microspheres

The students had done a simple experiment. They dumped a bunch of tiny spheres, known as "microspheres," into a beaker of water. They shook the suspension to ensure proper mixing, covered the beaker to minimize evaporation, and then went home for a good night's sleep. The next morning, they returned to examine the result.

By conventional thinking, nothing much should have happened, besides possibly some settling at the bottom of the beaker. The suspension should have looked uniformly cloudy, as if you'd poured some droplets of milk into water and shaken it vigorously.

The suspension did look uniformly cloudy — for the most part. However, near the center of the beaker (looking down from the top), a clear cylinder running from top to bottom had inexplicably formed (**Fig. 1.3**). Clarity meant that the cylinder contained no microspheres. Some mysterious force had driven the microspheres out of a central core and toward the beaker's periphery. If you've ever seen *2001: A Space Odyssey*, and the astonishment of the ape-humans upon first seeing the perfect monolith, you have some sense of just how our jaws dropped. This was something to behold.

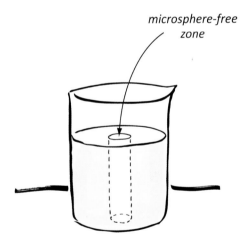

microsphere-free zone

Fig. 1.3 *Near-central clear zone in microsphere suspension. Why does the microsphere-free cylinder appear spontaneously?*

So long as the initial conditions remained within a well-defined window, these clear cylinders showed up consistently; we could produce them again and again.[2] The question: what drives the counterintuitive migration of the spheres away from the center? (Chapter 9 explains.)

(ii) The Bridge Made of Water

Another curious laboratory phenomenon, the so-called "water bridge," connects water across a gap between two glass beakers — if you can imagine. Although the water bridge is a century-old curiosity, Elmar Fuchs and his colleagues pioneered a modern incarnation that has aroused interest worldwide.

The demonstration starts by filling the two beakers almost to their brims with water and then placing them side-by-side, lips touching. An electrode immersed in each beaker imposes a potential difference on the order of 10 kV. Immediately, water in one beaker jumps to the rim and bridges across to the other beaker. Once the bridge forms, the two beakers may be slowly separated. The bridge doesn't break; it continues to elongate, spanning the gap between beakers even when the lips separate by as much as several centimeters (**Fig. 1.4**).

Astonishingly, the water-bridge hardly droops; it exhibits an almost ice-like rigidity, even though the experiment is carried out at room temperature.

I caution you to resist the temptation to repeat this high-voltage experiment unless you consider yourself immune to electrocution. Better to watch a video of this eye-popping phenomenon.[w1] The question: what sustains the bridge made of water? (See Chapter 17.)

(iii) The Floating Water Droplet

Water should mix instantly with water. However, if you release water droplets from a narrow tube positioned just above a dish of water, those droplets will often float on the water surface for a period of time before dissolving (**Fig. 1.5**). Sometimes the droplets may sustain themselves for up to tens of seconds. Even more paradoxically, droplets don't dissolve as single unitary events; they dissolve in a succession of squirts into the pool beneath.[3] Their dissolution resembles a programmed dance.

Fig. 1.4 *The water-bridge. A bridge made of water spans the gap between two water-filled beakers. What sustains the bridge?*

Fig. 1.5 *Water droplets persist on water surface for some time. Why?*

Floating water droplets can be seen in nature if you know where to look. A good time is just after a rainfall, when water drips from a ledge onto a puddle or from a sailboat's gunwales onto the lake beneath. Even raindrops will sometimes float as they hit ground water directly. The obvious question: if water mixes naturally with water, then what feature might delay the natural coalescence? (Look at Chapters 13 and 16)

(iv) Lord Kelvin's Discharge

Finally, **Fig. 1.6** depicts another head-scratching observation. Water drawn from an upside-down bottle or an ordinary tap is split into two branches. Droplets fall from each branch, passing through metal rings as they descend into metallic containers. The rings and containers are cross-connected with electrical wires, as shown. Metal spheres project toward one another from each container through metallic posts, leaving an air gap of several millimeters between the spheres.

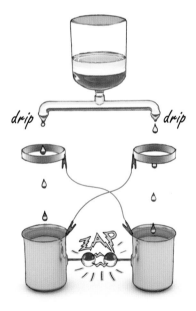

Fig. 1.6 *The Kelvin water-dropper demonstration. Rising water levels create a high-voltage discharge. Why does this happen?*

Originally conceived by Lord Kelvin, this experiment produces a surprising result. Once enough droplets have descended, you begin hearing a crackling sound. Then, soon after, a flash of lightning discharges across the gap, accompanied by an audible crack.

Electrical discharge can occur only if a large difference in electrical potential builds between the two containers. That potential difference can easily reach 100,000 volts, depending on gap size. Yet, the massive separation of charge needed to create that potential difference builds from a *single source of water.*

Constructing one of these exotic devices at home is possible[w2]; however, observing the discharge on video is a lot simpler. A fine example is the one produced by Professor Walter Lewin,[w3] who demonstrates the discharge to a classroom full of awe-struck MIT freshmen. He then invites the students to explain the phenomenon as their homework assignment. Can you explain how a single source of water can yield this massive charge separation? (Read about it in Chapter 15.)

Lessons Learned from These Mysteries

The phenomena presented in the foregoing sections defy easy explanation. Even prominent water scientists I know cannot come up with satisfying answers; most cannot get beyond the most superficial explanations. Something is evidently missing from our framework of understanding; otherwise, the phenomena should be readily explainable — but they are not.

I want to reemphasize that we're not dealing with water at the molecular level; we're dealing with crowds of water molecules. We don't yet understand water molecules' interaction with other water molecules — water's "social" behavior.

Social behavior is the purview of social scientists and clinicians, from whom we might learn. A friend of mine, a psychiatrist, once told me that, in order to understand human behavior, you should focus on oddballs and weirdos. Their behavioral extremes, the psychiatrist opined, provide clues for understanding the subtler behaviors of the rest of the population. That same reasoning can apply here: the foregoing cases describe some situations where water exhibits extreme "social" behaviors; as such, they provide clues for understanding the more ordinary behaviors of water molecules.

Thus, rather than brushing aside our inability to explain the phenomena above, we exploit them for the clues they provide. We turn ignorance to advantage. You'll see many examples of this process once we reach the book's middle chapters.

The next chapter provides some helpful background. It considers what we already know about water's social behavior and what we don't, but it focuses mainly on the surprising reasons why we know so little about Earth's most common substance.

2 The Social Behavior of H₂O

Water is central to life — so central that Albert Szent-Györgyi, the father of modern biochemistry, once opined: "Life is water dancing to the tune of solids." Without that dance, there could be no life.

Given that centrality, you might assume that we in the 21st century know pretty much all there is to know about water. All answers should be in by now. Yet the previous chapter confirmed otherwise, showing how little we really know about this familiar and pervasive substance.

Consider what Philip Ball has to say on that issue. Ball is one of the premier science writers of our time, author of *H₂O: A Biography of Water*, and a long-time science consultant for the journal *Nature*. Ball puts it this way[1]: "No one really understands water. It's embarrassing to admit it, but the stuff that covers two-thirds of our planet is still a mystery. Worse, the more we look, the more the problems accumulate: new techniques probing deeper into the molecular architecture of liquid water are throwing up more puzzles."

The water molecule itself is pretty well understood. Gay-Lussac and von Humboldt defined its essential nature just over two centuries ago; by now, fine details of its architecture are known. Essentially, the water molecule consists of two hydrogen atoms and one oxygen atom, arranged in a configuration that you might have seen in textbooks (**Fig. 2.1**).

We still know too little about how that molecule interacts with other water molecules or with molecules of different kind. Non-experts rarely raise questions of this nature. For most, it suffices to know that water molecules somehow link up with other water molecules. That's it. Biologists, for example, often regard water as the vast molecular sea that bathes the important molecules of life. We do not picture water molecules as seriously interacting with anything.

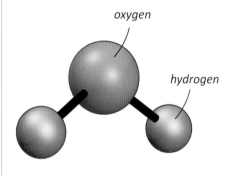

Fig. 2.1 *Artist's sketch of the water molecule.*

But water molecules must interact. Think of the simple water droplet: at least some of the gazillions of water molecules that make up the droplet must stick to others, for without cohesion there could be no droplet. Those cohesive interactions cannot be static. They must change as two droplets coalesce, and they must change as a droplet spreads on a surface. Even the simple droplet can't be understood without understanding water-water interactions.

So we ask, what is the nature of those interactions?

The Current Status of Understanding

Although a hodgepodge of ideas, the following list provides a short description of recent attempts to account for water's behavior. The theories of water-water interactions are complex, and even water scientists occasionally have difficulty understanding one another's theories. So, I will keep it brief. Readers seeking a more comprehensive understanding might find it useful to read a detailed review by Philip Ball.[2] Here I merely outline how seven prominent scientific groups think water molecules interact with one another (**Fig. 2.2**).

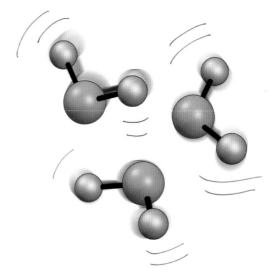

Fig. 2.2 *Interaction among water molecules. The nature of the interaction is not well understood.*

• The classical view of water-water interaction is the "flickering cluster" model introduced in 1957 by Frank and Wen. In this model, clusters of water molecules build from surrounding water. Positive feedback makes the clusters grow to a critical size and then spontaneously disperse. All of this happens on a time scale of 10^{-10} to 10^{-11} seconds; hence, the clusters "flicker." Although outdated, this model still appears in many textbooks.

• Martin Chaplin of London South Bank University, England, presents a model with slightly more organization. Chaplin suggests that liquid water consists of two types of intermixed nanoclusters. One type is empty, shell-like, and more-or-less collapsed, while the other is rather solid and more regularly structured. Molecules of water switch their allegiance rapidly between these two phases, but under a given set of conditions, the average number of molecules in each category remains the same. Those interested in this model can find details, and much more about water, on Chaplin's famously informative website.[w1]

• Quite a different picture emerges from the work of Anders Nilsson of Stanford University and Lars Petterson of Stockholm University. Their model also posits two coexisting types of water: ice-like clumps or chains containing up to about 100 molecules; and a disordered type of organization that surrounds those clumps. The authors envisage a kind of disordered sea, containing rings and chains of hydrogen and oxygen atoms.

• The model of Emilio del Giudice of the University of Milan is characterized by a much larger scale of clustering. Based on quantum-field theory, del Giudice posits submicron-sized coherence domains of water, each of which may contain many millions of molecules. The bonds between the water molecules within those domains may be thought of as antennae that receive electromagnetic energy from outside. With such energy, the water molecules can release electrons, making them available for chemical reactions.

• A popular model that builds on the associations inherent in all of the foregoing models comes from Gene Stanley of Boston University. Stanley suggests that water has two distinct states, low density and high density. The distinction appears most clearly in supercooled water. Low-density water has an open tetrahedral structure, while high-density water has a more compact structure. The two states dynamically interchange with one another.

• Another two-state model emphasizes that water molecules can exist as mirror images. That is, one fraction of water molecules is left-handed, while the other is right-handed. Major proponents of this kind of model include Sergey Pershin from Russia and Meir Shinitzky and Yosi Scolnik from Israel. They argue that the relative proportions of these two species can explain diverse features of water.

• The most structurally complex model, put forth by the late materials-science pioneer Rustum Roy, emphasizes the heterogeneity of water structure, as well as the ease of water-molecule interchange. Interchanges require very little energy. **Figure 2.3** shows a cartoon schematizing some representative structures.

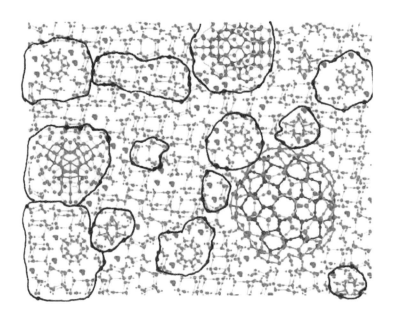

Fig. 2.3 *Proposed structure of liquid water, from Rustum Roy and colleagues.[3] Clusters are outlined in black.*

By now, you may feel you have heard enough about structural models. Yet this sampling is merely representative of a larger group of models that are continually argued and debated. Our understanding of water remains unresolved or, as Ball puts it, "a mystery."

On the other hand, most of these models share a common feature: multiple states. The common view is that liquid water has but one state; yet these models theorize some additional state. Later, we will see concrete evidence for a robust state of water that is visually detectable and endowed with well-defined features.

Why We Understand So Little

You might find it hard to believe, but few scientists study water. Most scientists presume, as do lay people, that everything about this common substance must already be known — so where's the scientific challenge? Better to pursue some trendy area like molecular biology or nanoscience rather than plunge into boring water.

Scientists shun water for a second reason. Water seems to have acquired a rather mystical character. Ancient religious gurus felt certain that water was endowed with exotic healing powers. Think of "holy water." This mystical tinge makes water research a potentially risky business: an exotic finding may be viewed as the work of the devil, rather than as the work of science. Better to avoid the risk of condemnation.

Despite those two disincentives, water once occupied a central position in scientific research. During the first half of the 20th century, science had a different emphasis than it does now. Rather than adding detailed knowledge to narrowly focused areas, scientists sought to uncover general principles that might apply throughout nature. The whole seemed more important than its molecular parts. That whole had to include water because water was virtually everywhere.

It was also a time when colloids, submicroscopic particles suspended in a liquid, seemed important. Believing that a colloidal foundation was the basis of life, many scientists assumed that knowledge about colloid-water interaction would elucidate life's underlying chemistry. The focus on colloids, combined with the holistic approach, put water at the center of scientific research.

But, by the middle of the 20th century, two things blighted the promising water harvest. First was the shift toward specialization. That shift drew scientists toward more molecular approaches that assigned water a secondary role. Molecules became the rage. The more you understood a molecule, it seemed, the closer you approached scientific truth. Inevitably, water research became old-fashioned and gradually lost its prominence.

The second thing that made scientists shy away from water involved two sociopolitical incidents, each of which had a terrible dampening effect on progress in understanding water.

The first incident, the so-called "polywater debacle," began during the Cold War, in the late 1960s, with a provocative Russian discovery. Water confined within narrow capillary tubes seemed to behave differently than ordinary water: its molecules vibrated differently; its density was anomalously high; and it was difficult to freeze or to vaporize. Clearly, this was some exotic brand of water. Because its properties implied the high stability common to many polymers, chemists thought of it as polymer-water and coined the ultimately fateful descriptor, "polywater."

The discovery of polywater triggered excitement among many scientists — imagine, a new phase of water. But the discovery also met with skepticism, and the Russians eventually wound up embarrassed when Western scientists identified an insidious problem: impurities. The supposedly pure water situated inside those capillary tubes was shown to contain salts and silica leached from the surrounding glass tubes. Those impurities had apparently given rise to the exotic features that were reported. Even Boris Derjaguin, the legendary physical chemist responsible for most of the initial studies, eventually admitted publicly that the impurities had been present. The skeptics could find justification in their initial reaction that polywater was "hard to swallow."

I'll have more to say later on polywater. I will just mention here that "contaminants" are bugaboos that plague all scientific fields. A scientist hopes for something pure, but absolute purity is often difficult to attain. In the case of water, achieving purity is virtually impossible because water has a propensity to absorb all kinds of foreign molecules; it's a natural solvent for almost everything. In this sense, contaminants are natural features of water, and their presence in limited quantities does not necessarily imply that any observed feature needs to be reflexively discarded.

However, the damage was done. By the early 1970s, the Russians were deemed guilty of careless experimentation. The injury to the field grew far out of proportion to the indictment's significance, mainly because of the sensational publicity given to polywater when the press caught hold of the story. Imagine, they suggested: a drop of polywater thrown into the sea could act like any polymeric catalyst — that single drop could polymerize the earth's entire water supply into a single blobby mass, which would end all life. Dangerous stuff, for sure (**Fig. 2.4**).

The public was therefore relieved by the reports of the contamination error. Other, less paranoid folks felt disappointed that this exciting new scientific finding turned out to be nothing more than an experimental flub. Either way, water scientists were considered incompetent.

The ensuing catastrophic impact on all water research is not difficult to imagine. If Russia's premier physical chemist could go so easily astray, then what about ordinary scientists? The risk of embarrassment seemed high. Talented scientists who might have pursued water research chose to work on safer subjects to avoid any possible taint of polywater.

So, largely out of fear, water research screeched to a halt. A few brave diehards persisted, mainly in the area of biological water, but the momentum was killed. The lingering mystery of water was left for others to resolve — sometime in the vaguely distant future.

Fig. 2.4 *The specter of polywater.*

The Water Memory Debacle

Two decades later, water science showed signs of incipient recovery — until an even deadlier blow struck: the so-called "water memory" debacle. Here, the central figure was the late French scientist and renowned immunologist, Jacques Benveniste. Almost by accident, Benveniste and colleagues obtained evidence that water could retain information from the molecules with which it interacted. Water, you might say, could "remember."

The evidence for water memory came from experiments involving successive dilutions of biologically active substances. Take such a substance dissolved in water, and dilute it. Then take a bit of this dilute solution and dilute it again; repeat this process again and again. After you have diluted it enough times, all you have left is water;

statistically, none of the original substance remains. Benveniste and colleagues would continue to dilute it even well beyond that stage of nothing remaining and still found that the solution could have as much biological impact as the original. Pouring either the concentrated substance or the serially diluted substance onto cells could trigger the same molecular dance. It appeared that the diluted water retained a "memory" of the molecules with which it had been in contact, for only those molecules were specific enough to initiate that dance.

Preposterous, thought the editor of *Nature*, Sir John Maddox. How on earth could water retain information? But not everyone shared that seemingly obvious response. Homeopaths use a similar procedure when preparing their remedies, and some members of the homeopathic community felt that a distinguished scientist had finally vindicated their approach. Benveniste, on the other hand, was less interested in homeopathy than in science. Reacting to the summary rejection of his findings by *Nature*, Benveniste asked colleagues in three other laboratories to repeat his experimental protocols to see if they could obtain the same results.

Remarkably, they did. And, once again, Benveniste submitted a report of the findings to *Nature*. The journal responded the same as before. Evidently, no matter how many laboratories could reproduce the result, the findings looked so improbable that some experimental gremlin clearly must have been lurking in that diluted water. With the polywater incident still very much in mind, *Nature* smelled a rat.

Under pressure to act fairly, the journal finally agreed to publish the results, albeit with one condition: the editor reserved the right to summon a committee to look over the shoulders of the French scientists as they performed their experiments; then the committee would report back to the readers of *Nature*. The French group accepted the stipulation. The paper quickly appeared, along with an appended disclaimer of skepticism. The editor indicated that he would launch an investigation: a committee of peers would determine just what those French scientists were really up to.

The committee of peers was, in fact, a committee of sleuths. Editor Maddox headed the committee. Maddox recruited two additional people. The first was Walter Stewart, who worked at the US National Institutes of Health in a special division dedicated to uncovering scientific fraud. Stewart was a professional sleuth. The other was James Randi, otherwise known as "The Amazing Randi." A world-class stage magician, Randi earned his fame by debunking the tricks of other magicians, such as Uri Geller's claim that he could levitate. Judging from the makeup of this committee of "peers," it was clear that Maddox suspected more than just an innocent error.

The committee came to Paris and carefully watched the experiments. The first sets of experiments went pretty much as claimed, and the French seemed to prevail in the early rounds. But when one of the visitors himself performed the dilutions, the results did not go as well. The visitors then huddled. They quickly concluded that, since the French could produce the claimed result but the visitors could not, therefore a trick must be at play. The nature of the trick remained unclear to the professional debunkers. Nevertheless, their report to the world of science boldly declared that water memory was "a delusion."

This colorful story is rich with detail, and for more of that I recommend two books. The first is the above-cited book by Philip Ball,[1] who worked for *Nature* at the time and was close to Maddox. The second book, entitled *The Memory of Water,*[4] was written by the late physicist Michel Schiff. Schiff had been working in the French laboratory at the time of the incident. As you may imagine, these authors have rather different sympathies. To get the full picture, you should read both books.

As a result of this fiasco, Benveniste suffered widespread humiliation. That humiliation included the loss of grant support, the collapse of a large and productive laboratory, difficulty publishing any further scientific work, and — the ultimate ignominy — twice winning the "Ig-Nobel" Prize, awarded by Harvard students for improbable research. It was not a happy time for French science (**Fig. 2.5**).

The main point, however, is neither the ugliness of the incident nor the instant demise of an illustrious scientific career; the main point is the impact this had on the field of water research. Barely having recovered from the polywater debacle, the field suffered this second, even more devastating setback. Water memory became the laughingstock of the entire scientific community. Finding it hard to remember names? Try drinking more water. (Ha, ha!)

Fig. 2.5 *An embarrassment to French science?*

Given this troubled history, you can imagine the consequence for water research. How many scientists of sound mind would dare enter a field first tainted by polywater and then debased as the butt of scientific jokes? Very few indeed. Yet there is some irony, for others would later confirm Benveniste's result,[5] and still others, including Nobel laureate Luc Montagnier, would build on water memory to claim transmission of information stored in water.[6] Despite all that, water memory remains largely a joking matter rather than a subject of serious scientific investigation.

The Mystery Lingers

I think you can now appreciate the paradox: why we have come to know so little about something so familiar. Two successive debacles have turned a once-dynamic field into a treacherous domain into which few scientists have the temerity to enter.

Rising from the ashes of those two debacles is the current field of water research. The field may be best described as schizophrenic. On the one side, mainstream scientists employ computer simulations and technologically sophisticated approaches to learn more about water molecules and their immediate neighbors. Their results more or less define the field. Taking relatively risk-free approaches, they have provided incremental advances that help refine and embellish the various models outlined earlier in this chapter.

On the other side are the scientists who explore the more provocative phenomena, such as those described in the previous chapter. The very mention of those phenomena often provokes a chuckle from mainstreamers, who consider the phenomena odd and less than scientific. Some mainstreamers like to dismiss those phenomena as a species of "weird water."

Rarely do the two sides mix. The weird water folks admire the mainstreamers' sophistication but often find their approaches dense and impenetrable; hence, they keep their distance. Mainstreamers, in turn, avoid the weird water folks like the plague. Some mainstreamers cringe at the prospect of yet another water debacle. Weird-water phenomena are thus consigned to fringe science — placed in the same category as cold fusion, UFOs, and subtle energies. You'd better keep your distance if you hope to retain your scientific respectability.

Given this atmosphere of suspicion, you can appreciate why building understanding has become a challenge. Conducting fundamental research on water is something like searching for gold nuggets in the mud. A few can be found here and there, but this slow, arduous gathering process occurs in an atmosphere of suspicion that makes it impractical to lay even a primitive foundation of understanding.

. . .

The chapters that follow will bypass this muddy, well-trodden pathway. We will forge an entirely fresh trail built on clues that others have ignored, and use this path to progress toward a better understanding. We take the position that the social behavior of water should not be as incomprehensible as now conceived: if nature itself is simple and intuitive as many scientists think, then we'd hope that its most ubiquitous component might be equally simple and intuitive.

It is this simple understanding that we strive to uncover.

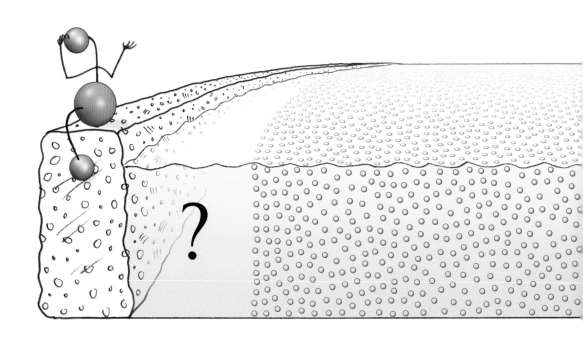

3 The Enigma of Interfacial Water

In a glass, all of the water looks the same. Peering intently into the glass provides no hint that molecules in one region might arrange themselves differently from molecules in another region. After all, water is water.

On the other hand, superficial appearances can deceive. I learned only during the past decade that material surfaces can profoundly impact nearby water molecules — so profoundly and extensively that most everything about that water radically changes. Practically any surface that touches the water will have such effects: the container, suspended particles, or even dissolved molecules. Surfaces of all kind profoundly affect nearby water molecules.

Had I bothered to read the literature, I would have been fully aware of this surface impact: a half-century-old review article by JC Henniker[1] cites more than a hundred published studies confirming the long-range effect of varied surfaces on many liquids, including water. The evidence has been widely available.

For me, however, such long-range effects were a fresh revelation. I had been aware of surfaces affecting water out to perhaps tens of water molecule layers; I had even written a book on the biological relevance of such ordered water.[2] However, a truly long-range impact extending up to thousands or even millions of molecular layers was rather jarring. If true, this strong influence seemed inescapably central for all water-based phenomena.

I'll describe how we first stumbled upon evidence for this long-range ordering, and what we did to check that the evidence was sound. The alert came from a chance encounter at a scientific conference.

Lunch with Hirai

On a blisteringly hot summer day in the late 1990s, while darting from one building to another to attend a seminar, I had the good

Toshihiro Hirai, Shinshu University.

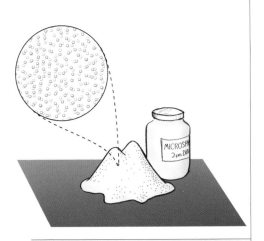

Fig. 3.1 *Microspheres are common tools for scientists.*

fortune to run into Professor Toshihiro Hirai from Japan's Shinshu University. We chatted at length. I described the book I was then writing on the role of water in cell function (*Cells, Gels, and the Engines of Life*). The subject evidently caught his attention, for as we proceeded to lunch to escape the heat, Hirai informed me of a seemingly relevant observation that his students had made — one that ultimately proved pivotal for understanding water.

Hirai and his students had been studying blood flow in vessels. In lieu of actual vessels, they used cylindrical tunnels bored through gels; for blood, they used suspensions of microspheres (**Fig. 3.1**). Thus, water suspensions of tiny spheres pumped through gel tunnels mimicked the blood flowing through vessels. The investigators could track the flowing "blood" because the gel was transparent; all they needed was a simple microscope.

Hirai eagerly shared their observations with me. I found his results on the patterns of blood flow illuminating, but what really caught my attention was his description of the odd behavior of the microspheres. He told me that the flowing microspheres avoided the annular zone just inside the gel surface; they restricted themselves to the tunnel's central core (**Fig. 3.2**). Hirai indicated that he did not pay particular

Fig. 3.2 *Schematic diagram illustrating the microsphere-free zone just inside the gel tunnel.*

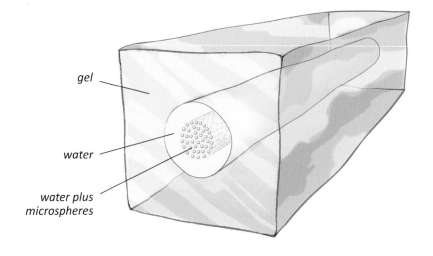

gel

water

water plus microspheres

attention to this feature, assuming it was a secondary effect. The possible centrality of this near-surface exclusion apparently had not occurred to him.

Following that encounter, Hirai and I exchanged many emails. Exercising care to avoid overstepping the boundaries of polite Japanese communication, I tried to persuade Hirai to publish his findings, as I had hoped to cite them in my then-forthcoming book. That was not to happen. Hirai grew justifiably impatient with my incessant emails and finally offered to include me as coauthor of any forthcoming publication while allowing him to proceed at his own pace.

To the best of my knowledge, Hirai's observations remain unpublished. However, quite serendipitously, a former postdoctoral research fellow of his moved to Seattle and walked into my lab looking for work. I instantly hired Jian-ming Zheng (**Fig. 3.3**), and we proceeded to follow up on Hirai's observations.

Fig. 3.3 *Jian-ming "Jim" Zheng.*

I had reason to suspect that the microspheres' inclination to avoid the zone near the gel surface might indicate something significant. It seemed possible that the gel surface might order contiguous water molecules; the growing order would then push out microspheres in the same way as growing ice crystals push out suspended debris. This hypothesis was unorthodox; however, my 2001 book detailed a substantial body of evidence pointing to that very notion.

The most astonishing aspect of Hirai's observations, however, was the scale. The microsphere-free zone extended about a tenth of a millimeter inward from the gel surface, implying that the ordered lineup might include hundreds of thousands of water molecules. That's akin to a lineup of marbles extending over several dozen US football fields. Even as an author championing the idea of water ordering in the cell,[2] I had trouble with that colossal magnitude; the span seemed too long.

I might have been a tad less skeptical had I been properly aware of the older scientific literature. Published over sixty years ago and based on numerous published papers, the review article that I mentioned[1] drew a similar conclusion: surfaces exert long-range influence on contiguous liquids; they bring substantial molecular reordering. Unaware of this evidence, we naïvely went on to reinvent the wheel.

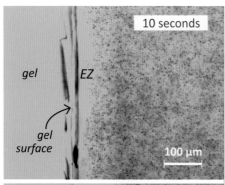

Fig. 3.4 *Microsphere-exclusion zone (EZ) next to a gel surface. The zone grows with time and then remains relatively stable after about five minutes.*

We started with simpler initial experiments than Hirai's. Using the same type of gel, we plunked a piece into a chamber and suffused it with an aqueous suspension of microspheres. We then looked into the microscope to see what might happen. As soon as the liquid suspension met the gel, the microspheres began moving away from the gel's surface, leaving a microsphere-free zone just under $100~\mu m$ (0.1 mm) wide. Water remained in that zone, but microspheres did not. Once formed, the zone remained intact: even after several hours of examination, the microspheres resisted invasion. **Figure 3.4** shows the development of this microsphere "exclusion zone."

Our observations revealed that the microsphere-free zone seen by Hirai did not arise from the hydrodynamics of "blood" flow; our setup had no flow, yet we obtained a similar zone of exclusion. Something about the gel surface appeared to drive the microspheres into hasty retreat — with or without imposed flow. Both scenarios produced the same result: a distinct exclusion zone, or "EZ," as we came to call it.

The Conventional Expectation

The exclusion phenomenon seems to fly in the face of the tenets of modern chemistry. The phenomenon should not exist. Surfaces may certainly affect the adjacent liquid, but it is widely presumed that the impact does not project into the liquid beyond a few molecular layers (despite the evidence cited in Henniker's review article).

Why so limited an impact? The prevailing view derives from the theorized presence of an electrical "double layer" of charge. Thus, a charged surface placed in water will attract oppositely charged ions dissolved in that water (**Fig. 3.5**, *opposite page*). Beyond that ion layer lies a second layer whose polarity is opposite the first, extending diffusely into the liquid. And beyond that double layer must lie additional diffuse charges, etc. Eventually, neutrality prevails. To an observer situated beyond those neutralizing layers, the surface should be unnoticeable — as though the surface were absent.

That minimum distance for insensitivity is labeled the "Debye length," after the Dutch physicist Peter Debye. The value of the Debye length reflects the extensiveness of the counter-ion clouds. Although the exact value depends on many factors, typical values are on the nanometer (10^{-9} meter) scale. Beyond those several nanometers, according to theory, any solute or particle situated in the liquid should be insensitive to the presence of the material surface.

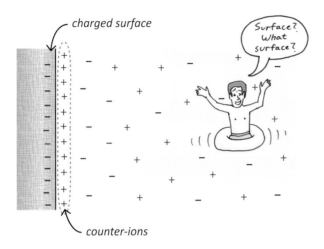

charged surface

counter-ions

Fig. 3.5 *Standard double-layer theory. Charged surface (left) is expected to attract counter-ions of opposite polarity, as shown. Those counter-ions then attract a diffuse cloud of opposite charges, etc. An observer sitting in the water at a site far from the interface should not sense the neutralized surface.*

That is not what we observed (**Fig. 3.4**). Particles were *markedly* sensitive to the material surface — distancing themselves from the surface by some 100,000 times the Debye length.

That observation spelled trouble, because the Debye length and double-layer theory are bedrock concepts of surface chemistry. Challenging that theory with conflicting experimental observation meant that we had to make certain; we had to be sure that no trivial explanation or underlying artifact (scientific jargon for error) might have confounded our observations.

Trivial Explanation?

Zheng and I dedicated a full year to probing every conceivable error.[3,4] We got lots of input from others, who were not shy about suggesting gremlins that might lurk insidiously beneath the interpretational surface. Of the many issues we addressed, four seemed particularly problematic.

• The first issue involved convectional flow that might arise from slightly different temperatures in different regions. Such temperature gradients might create fluidic swirls that could draw microspheres away from the surface. In many experiments, we did observe convectional flows; in other experiments, however, flow was altogether absent, and yet the exclusion zone persisted. We concluded that convectional flows could not provide a general explanation for the observed exclusion zones.

• A second issue was the polymer-brush effect. Gels are made of polymers (large molecules consisting of repeating structural units), whose strands might project beyond the gel proper and into the surrounding solution — like the bristles of a brush. Sparse, thin bristles might escape microscopic detection while excluding microspheres. However, running an ultrasensitive nanoprobe parallel to the gel surface revealed no evidence for any such bristles. The invisible bristle argument seemed bogus.

Subsequent experiments confirmed that conclusion. One of those experiments used self-assembled monolayers, i.e., single molecular layers functionalized with charge groups. Monolayers have no projecting polymers. Yet they could produce exclusion zones of ample size.[4] We also saw substantial exclusion zones next to certain n-type silicon wafers, as well as next to metal surfaces,[5] which, again, contain no projecting bristles. **Figure 3.6** shows an example.

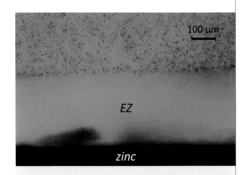

Fig. 3.6 *Exclusion zone next to zinc, from reference 5. Green color results from using a green filter in the microscope.*

• A third trivial explanation for microsphere exclusion invoked long-range electrostatic repulsion. If both the material surface and the microspheres are negatively charged, then the two entities should repel; strong enough repulsion should drive away the microspheres, creating a zone of exclusion. We considered this hypothesis even though double-layer theory predicts that any such repulsion ought to vanish at separations beyond a few nanometers, a distance some 100,000 times smaller than what we regularly observed.

The simplest test of the repulsion hypothesis was to substitute positive microspheres for negative microspheres. According to the electrostatic hypothesis, the positive microspheres should be drawn toward the negative surface. We found that the positively charged microspheres did sometimes collapse the exclusion zone; in other instances, the exclusion zone not only remained, but also remained the same size as seen with the negative microspheres.[3,4]

We got a similar result when we reversed the charge of the excluding surface. For those experiments, we used gel beads, whose spherical surfaces create shell-like exclusion zones (**Fig. 3.7**). Negatively charged microspheres were consistently excluded. It didn't matter whether the beads' surface contained negatively charged or positively charged polymers.[6] Simple electrostatic repulsion cannot explain these results.

• A fourth possibility involved some material diffusing from the gel. Leaking contaminants might conceivably push away the microspheres, leaving an apparent zone of exclusion. However, monolayer results contradict that hypothesis: those single molecular layers produced substantial exclusion zones,[4] yet they are so thin that virtually nothing is available to leak out.

• We also tried another approach: washing away any putative leaking contaminants. Vigorous flow parallel to the EZ-nucleating surface, no matter how swift, could not eliminate the EZ.[7]

• Finally, we could find exclusion zones too extensive to be explained away by leaking materials. Such extensive EZs were found in long, horizontally oriented cylindrical chambers. At one end of the cylinder, we mounted a disc-like gel held by clips. We then filled the chamber with a microsphere suspension and watched. A pancake-like exclusion zone grew, as expected, from the gel surface to a thickness of several hundred micrometers. But the growth didn't stop there (**Fig. 3.8**); the EZ continued to grow by wedging down to pole-like projections. Sometimes branching, those pole-like EZs typically extended to the very ends of meter-long chambers.[8] Clearly, a diffusing contaminant could not account for these ultralong exclusion zones.

Our yearlong studies lent confidence that the observed exclusion zones do not arise from trivial explanations. At this writing, several dozen laboratories have confirmed the existence of EZs. Furthermore (and to our chagrin), a recently uncovered paper published in 1970 showed largely the same results: microsphere-excluding zones several hundred micrometers thick, found adjacent to polymeric and biological gel surfaces.[9] Hence, microsphere exclusion is not a fluke. Something unpredicted is happening that drives microspheres from certain material surfaces.

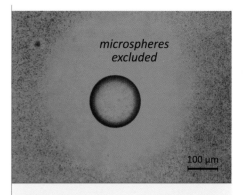

Fig. 3.7 *Microspheres excluded from the vicinity of a charged gel bead, as seen in an optical microscope. (Color arises because of microscope filter.) We positioned the bead on a glass surface and added the microsphere suspension. The EZ grew with time to the extent shown.*

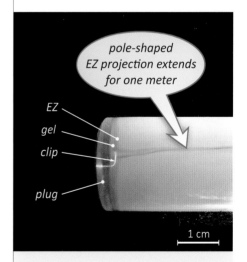

Fig. 3.8 *Long EZ projection. The disc-like gel creates a disc-like EZ that wedges into a long pole-like projection. The projection can extend at least one meter.*

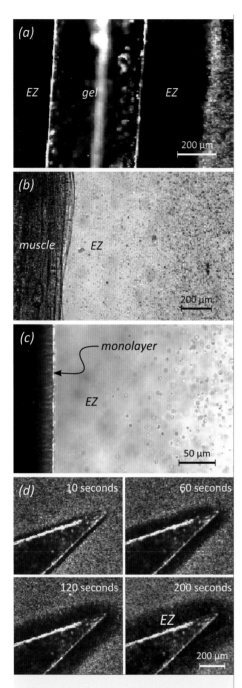

Fig. 3.9 *Examples of microsphere-excluding zones, viewed in an optical microscope. (a) polyacrylic acid gel; (b) muscle; (c) a self-assembled monolayer on gold. (d) Nafion polymer, time series.*

Although our artifact-seeking experiments consumed a good deal of our energy, they brought an unexpected clue. Those meter-long exclusion zones struck us as implying some kind of crystal-like structure, for crystals easily grow to such lengths: think of an icicle. Crystals also exclude particles as they grow. The prospect that the EZ might be some kind of crystal-like material intrigued us.

Crystals generally grow from nucleation sites, i.e., from surfaces of some kind. It seemed important therefore to determine what kinds of surfaces nucleate exclusion zones.

How General Are Exclusion Zones?

We first examined several gels over and above those mentioned. All water-containing (hydro)gels produced exclusion zones, including gels made of biological molecules and artificial polymers (**Fig. 3.9a**). We also saw exclusion zones next to natural biological surfaces; they included vascular endothelia (the insides of blood vessels), regions of plant roots, and muscle (**Fig. 3.9b**). I already mentioned monolayers (**Fig. 3.9c**). Seeing substantial EZs adjacent to single molecular layers told us that material depth was not consequential: it appeared possible that *creating an exclusion zone merely required a molecular template.*

Various charged polymers also produced exclusion zones. An especially potent one was Nafion (**Fig. 3.9d**). Nafion's Teflon-like backbone contains many negatively charged sulfonic acid groups, which make this polymer one of the more potent excluders. Because of Nafion's robust exclusion zones and ease of use, you'll see it mentioned frequently in these pages.

The only exotic features we encountered were breaches — localized surface patches devoid of EZs. Those bare patches were atypical. However, they could be found regularly next to certain metals, and also next to polymeric membranes when straddled by differing solutions, as was the case in our osmosis experiments (see Chapter 11). Those EZ breaches seemed rather like holes penetrating through the ordinary EZ dam.

The EZ-nucleating materials described in the paragraphs above fall into the category of "hydrophilic," or water loving. Their love for water seems profound enough to exclude other suitors; only the water gets to stay. "Hydrophobic" or water-hating surfaces, such as Teflon, prove inept by contrast; no exclusion zones could be found. It appears that *the exclusion phenomenon belongs to hydrophilic surfaces as a class.*

Having established the EZ's generality, we next asked: what does the EZ exclude? Does it exclude microspheres alone? Or are other substances excluded as well?

We found a wealth of excluded substances, ranging from large suspended particles down to small dissolved solutes.[3] Microspheres of all kinds were excluded. They ranged in size from 10 μm down to 0.1 μm and were fabricated from diverse substances. Even red blood cells, several strains of bacteria, and ordinary dirt particles scraped from outside our laboratory were excluded. The protein albumin was excluded, as were various dyes with molecular weights as low as 100 daltons — only a little larger than common salt molecules. The span between the largest to smallest of the excluded substances amounted to a thousand billion times (**Fig. 3.10**).

These experiments showed that the EZ rather broadly excludes substances of many sizes, from very small to very large.

We could not definitively test the tiniest of solutes — that had to wait. Nevertheless, we could conclude that the exclusion phenomenon was general: *almost any hydrophilic surface can generate an EZ, and the EZ excludes almost anything suspended or dissolved in the water.*

Why Are Solutes Excluded?

This demonstrably vast exclusionary power implied yet again that we might be dealing with some kind of crystal-like substance, for crystals exclude massively. I alluded earlier to a possible crystalline structure: the hydrophilic surface could induce nearby water molecules to line up as they would in a liquid crystal. As the ordered zone grew, it would push out solutes in the same way that a growing glacier pushes out rocks.

Such molecular ordering is not a new idea. The previously referenced Henniker paper (1949) reviews many older works showing massive near-surface molecular reordering. Henniker's was not a voice lost

one

one thousand billion

Fig. 3.10 *Range of excluded substances.*

Fig. 3.11 *Albert Szent-Györgyi in his later years.*

Fig. 3.12 *Gilbert Ling in his earlier years.*

in the wind. Subsequently, the idea of long-range water ordering was advanced by a number of prominent scientists, including Walter Drost-Hansen, James Clegg, and especially Albert Szent-Györgyi and Gilbert Ling. Szent-Györgyi (**Fig. 3.11**) was a seminal thinker who won the Nobel Prize for discovering vitamin C. A cornerstone of his thinking was the long-range ordering of water, which he regarded as a major pillar in the edifice of life.

Gilbert Ling (**Fig. 3.12**) thought similarly. He emphasized the central role of water ordering in cell function, building a revolutionary framework for biological understanding. He wrote five books on this subject, the latest being his 2001 monograph, *Life at the Cell and Below-Cell Level.*[10] This book argues that the cell's charged surfaces order nearby water molecules, which in turn exclude most solutes. According to Ling, this ordering is the very reason why most solutes occur in low concentrations inside the cell: the cell's ordered water excludes them.

With the stage amply set by these towering figures, the idea that charged or hydrophilic surfaces might order water molecules out to appreciable distances seemed plausible; we found solid experimental precedent. It was also clear that today's mainstream chemists thought this kind of ordering unlikely because molecules tend toward disorder. Nevertheless, some mechanism had to explain the profound exclusion, and water ordering seemed a viable option. Our lab therefore set out to explore that possibility.

Additional Evidence that Surfaces Impact Nearby Water

To determine the physical nature of the exclusion zone, we pursued a variety of methods. In each, we set up an exclusion zone (always using the purest water obtainable); we tested whether the particular property under investigation in the exclusion zone differed from the water beyond the exclusion zone. By doing so, we hoped not only to test for a difference, but also, if we were lucky, to pin down the nature of EZ water. What follows is fairly technical, but I hope you will bear with me through the description of six important experimental tests.

(i) Light absorption. Substances differ in the way they absorb light. By charting the absorption of differing wavelengths ("colors"), we

learn how a substance accepts electromagnetic energy; this can tell us how the molecules deal with that absorbed energy. At the very least, we hoped to see whether the wavelengths of light absorbed by the EZ differed from the wavelengths absorbed by the bulk water beyond.

To test for such differences, we set up the experiment shown in **Fig. 3.13a**.

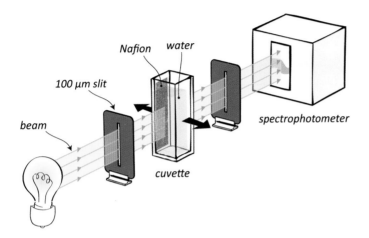

We bonded a sheet of Nafion to the inside face of a standard optical container, or cuvette, which we then filled with water. As the figure shows, we placed the cuvette in the path of a narrow window of light that would penetrate the water before reaching the spectrophotometer; moving the cuvette in measured increments let us investigate the light passing through regions both within and beyond the EZ.

Figure 3.13b shows the results. Far from the Nafion-water interface (beyond 400 μm), the spectrum was flat — i.e., the absorbed wavelengths of visible and near-visible light were no different from a blank water sample with no excluding surface present. That was anticipated. However, shifting the cuvette so that the illuminated window came closer to the Nafion-water interface and within the EZ caused a strong absorption peak to appear. Its wavelength was approximately 270 nm. The 270-nm absorption peak grew with the window's proximity to the Nafion surface and eventually dominated the absorption spectrum. Since no such peak appeared in the water beyond the EZ, it became clear that the absorption features of the EZ differ remarkably from those of the bulk-water zone.

Fig. 3.13a *Measurement of light absorption. Moving the cuvette laterally allowed us to interrogate water at various distances from the Nafion surface.*

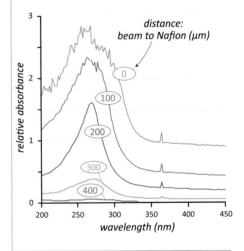

Fig. 3.13b *Absorption spectrum measured at various distances from the Nafion-water interface. Decreasing distances range from green to red. Numbers attached to each curve denote actual distances.*

(ii) Infrared absorption. Absorption differences can also be tested in the infrared region of the electromagnetic spectrum. Those longer wavelengths tell us something about molecular structure. **Figure 3.14** shows one result, a map of infrared absorption in and around a submerged triangular piece of Nafion. The different colors indicate different absorption magnitudes. Far from the Nafion, the uniform blue color indicates a uniformly low level of absorption. The color change closer to the Nafion (green) indicates that the EZ absorption differs from bulk water absorption.

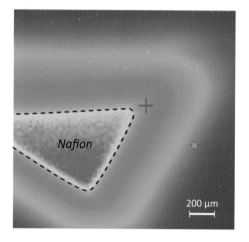

Fig. 3.14 *Triangular specimen of Nafion in water examined using infrared absorption. Color differences indicate differences of absorption. Blue is lowest.*

More detailed information may ultimately come from using thinner samples, but appropriately thin samples are challenging to produce; hence, their use may require technical advances. Nevertheless, the absorption differences seen in the current figure indicate that bulk water's structure differs from EZ water's structure.

(iii) Infrared emission. Our third approach used an infrared camera to measure the infrared radiation ("heat") emitted from a specimen. If the EZ's character differs from that of bulk water, then we might expect some difference in radiant emission.

To make the emission measurement, we placed a piece of Nafion in a shallow chamber containing water. We allowed the specimen to equilibrate for one hour. We then collected infrared radiation from the sample and averaged the radiation over multiple image frames. **Figure 3.15** shows a representative result. The dark region adjacent to the Nafion is the exclusion zone; it is dark because it radiates very little. More distant water regions radiate more brightly.

Interpreting the result requires some understanding of what determines infrared intensity. Hotter substances radiate more infrared — that's how airport thermal-image scanners can detect whether you have the flu and whether you may need to be quarantined for a week instead of lounging on the beach. Temperature, however, does not uniquely determine infrared intensity: intensity is the product of temperature and "emissivity" — the latter indicating the character of the emitting structure. Ordered, crystal-like structures emit less infrared energy than disordered structures because a crystal's molecular components move around less vigorously; those components are more stable. Thus, the generation of less infrared energy could mean either more stability or lower temperature.

Lower temperature does not explain the EZ's lower infrared emission seen in **Figure 3.15**. The records were averaged over extended periods of time during the experiment, so any temperature difference between the EZ and the bulk-water zone should have vanished. Emissivity differences seem the more plausible explanation. The darker EZ implies lower emissivity; i.e., the EZ is more ordered and crystalline than bulk water.

(iv) Magnetic resonance imaging. Magnetic resonance imaging (MRI) is a technique used for imaging tumors. Raymond Damadian, the pioneer who patented the technique, based his invention on the principle that water's character differs in different environments; this permits spatial imaging. In our MRI experiment, we placed a gel and adjacent water in the test area. The MRI imparts a pulsed magnetic field that excites water's atomic nuclei, whose protons then relax back down to their ground states. The relaxation time yields information on the degree of motional restriction relative to nearby molecules. The MRI computer then reconstructs this restriction data to create an image.

Figure 3.16 shows a map of relaxation times. Darker regions denote shorter relaxation times, which means more restriction. The map shows a dark band across the middle; this band coincides with the width and location of the EZ. Apparently, molecules within the EZ suffer more restriction than the water molecules beyond that zone.

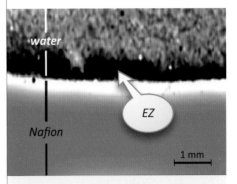

Fig. 3.15 *Infrared emission image of Nafion next to water. Sample was equilibrated at room temperature. Black band running horizontally across the middle of the image corresponds to the expected location of the exclusion zone.*

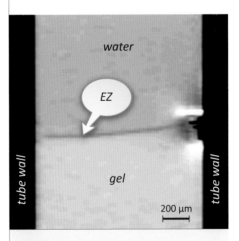

Fig. 3.16 *MRI map of relaxation times. The lower half of a capillary tube was filled with polyvinyl alcohol gel, while the upper half was filled with water. The dark band, corresponding to the gel's EZ, indicates more molecular restriction.*

This conclusion is not unique. An earlier study reported a similar restriction extending over even longer distances from material surfaces;[11] and a subsequent report from our laboratory[12] found that water near a surface exhibited a "chemical shift," which is jargon for implicating a different chemical species. Magnetic resonance techniques reveal substantial differences between EZ water and bulk water.

(v) Viscosity. We also measured viscosity, which reflects the degree of liquidity. Honey, for example, is more viscous than water. To test whether the viscosity of the EZ differs from that of bulk water, we used a technique called falling-ball viscometry. We lined the bottom of a small chamber with a sheet of Nafion and filled the chamber with water. Spheres of polymeric material were then dropped into the water. The spheres descended at a roughly constant velocity but progressively slowed as they entered the region of the exclusion zone **(Fig. 3.17)**. Speed reduction implies higher viscosity. This demonstrated that EZ water has a higher viscosity than bulk water.

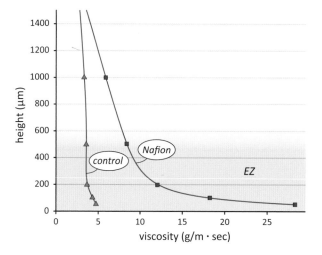

Fig. 3.17 *Viscous character of the EZ (shaded). We measured viscosity in water at various heights above a Nafion surface (red curve). Control (green curve) was obtained with a surface exhibiting little or no exclusion zone.*

(vi) Optical features. Two Russian groups independently measured the exclusion zone's refractive (light-bending) properties.[13,14] Both found that the EZ had a refractive index about 10 percent higher than that of bulk water. A higher refractive index ordinarily implies higher density; this suggests that EZ water is denser than bulk water.

All six sets of experiments — additional details of which are given elsewhere[4] — show that the water in *the exclusion zone differs in character from the water beyond the exclusion zone*. The differences are appreciable. EZ water is more viscous and more stable than bulk water; its molecular motions are more restricted; its light-absorption spectra differ in the UV-visible light range, as well as in the infrared range; and it has a higher refractive index. These multiple differences imply that EZ water fundamentally differs from bulk water. The EZ hardly resembles liquid water at all.

Order in the Exclusion Zone

To account for the nature of the EZ, our favored hypothesis was ordered water. The experimental results just considered seemed consistent with water ordering, but those experiments did not address the structural issue directly. For that, we needed other kinds of evidence.

We had good experimental reason to suspect order. Mae-wan Ho's wonderful book, *The Rainbow and the Worm*,[15] already adduced evidence for long-range order. Ho **(Fig. 3.18)** used a sensitive polarizing microscope. Polarizing microscopy is a standard method for detecting order, particularly in minerals. The principle is simple: if molecular structures line up, then the optical properties in the lined up direction will differ from those in orthogonal directions, giving rise to so-called birefringence. Ho shows structural lineups that extend over vast regions of a worm's body, concluding that the observed ordering comes largely from the ordering of water. **Figure 3.19** shows an image from her book.

Fig. 3.18 *Maewan Ho.*

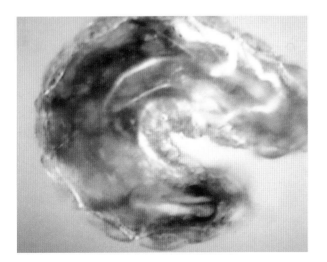

Fig. 3.19 *Freshly hatched Drosophila larva under the polarizing light microscope set up to optimize detection of liquid crystalline phases based on interference colors. The colors indicate that essentially all the molecules, including the water, are aligned; the particular colors depend on the orientation of the molecular alignment and their degree of birefringence. For more details, see Ho[15] pp. 219–221.*

Motivated by Ho to investigate this phenomenon, we set up our own polarizing microscopy system, which we used to explore water ordering in the vicinity of Nafion. Some experiments showed no clear birefringence, possibly because of insufficient sensitivity; other experiments gave positive results, which confirmed Ho's observations. **Figure 3.20** shows the water far from the Nafion interface as blue, indicating no preferred molecular orientation. Closer to the interface, the green color indicates a preferred molecular orientation. The ordered region corresponds to the zone of exclusion immediately adjacent to the Nafion. In other words, water in the exclusion zone is more ordered than the bulk water farther away.

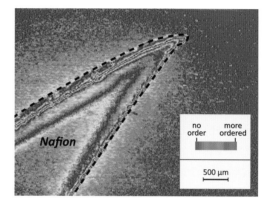

Fig. 3.20 *Arrowhead-shaped piece of Nafion sheet (delineated by broken line) in water, examined using polarizing microscopy. Blue color indicates a random orientation of molecules; red (see scale at right) indicates the highest degree of molecular ordering.*

The ordered zone in **Figure 3.20** is huge relative to water's molecular dimensions. Think of the water molecule's diminutive size: on the order of 0.25 to 0.3 nanometers (less than a millionth of a millimeter). The ordered zone in the figure corresponds to a lineup of approximately a million of those water molecules — like the lineup of marbles over dozens of football fields.

Two papers address the theoretical plausibility of such long-range ordering. One paper comes from the late Rustum Roy, a pioneer in the materials science field. Roy and his colleagues[16] stressed the precedent for certain surfaces to have a template-like effect, ordering molten materials into extensive crystalline arrays. Routinely used with semiconductor materials such as silicon, this process has made possible modern integrated circuits. It is also employed with molten aluminum. A similar process occurs during the formation of ordinary ice. Such precedents led Roy and his colleagues to suggest a similar template-based ordering of water molecules. They suggested that it was inevitable.

Arguing from a physicochemical point of view and from the results of numerous experiments, Ling[17] came to a similar conclusion: extensive ordering of water molecules nucleated from surfaces. Under ideal conditions, that ordering can extend to huge distances. That is, the proclivity to order can easily outweigh the natural tendency to disorder.

These two papers provide theoretical underpinnings for the molecular ordering we observed. They also offer a counterbalance against the commonly presumed impossibility of long-range ordering. On the other hand, unanswered questions remain. Neither the experimental evidence nor these theoretical considerations answers the questions: How exactly do the water molecules order themselves? Do water molecules merely stack? Or is some more elaborate type of reorganization at play? Answers to those questions will be coming next.

Reflections

I recognize that people nurtured on textbooks of modern chemistry may find little here that strikes a resonant chord. Textbooks imply something quite different from what we have found. Their emphasis on double-layer theory leads to the presumption that no more than a few layers of water molecules could possibly organize next to charged surfaces. Beyond those few layers, not much of note should be happening.

On the other hand, scientists have begun to recognize that water has properties not quite so mundane. Many water-based phenomena — a number of them considered in this book's opening chapter — have resisted explanation. Because of those difficulties, unsuspected features of water are now being considered more openly; i.e., the field has begun opening up to fresh and unexpected findings, one of which includes the long-range ordering of water.

Building on the evidence for long-range ordering, the next group of chapters uncovers an EZ structure surprisingly like ice. However, it is not ice. The ice-like ordering turns out to be the proverbial tip of the iceberg: something deeply consequential drives the buildup of ordered water in the EZ. That driving agent turns out to be a kind of energy common in everyday life and simple enough for anyone to understand.

The previous section showed that water situated near surfaces differs from bulk water. Next, we will explore the nature of that difference. We identify the structure of the near-surface exclusion zone and detail the features of that zone that will inform all that follows.

SECTION II

The Hidden Life of Water

4 A Fourth Phase of Water?

A s a college freshman back in the dark ages of 1957, I recall the stir created by the launch of the world's first space satellite. Sputnik was a stunning achievement. It was also a Soviet coup that caught the US off guard — which felt ominous during that tense Cold War era. Galvanized by that Soviet feat, the US government responded by massively funding scientific research and engineering development. Sputnik had been an embarrassment the likes of which the government could not allow to happen again.

But, only a decade later, a new embarrassment seemed imminent. This time, the source of the problem arose from something less lofty than a satellite: it came from water science. The Russians had struck again, having apparently discovered a new phase of water. Russian scientists had put some water into narrow capillary tubes and found that the water's properties changed dramatically. The water no longer behaved like a liquid; nor was it a solid. For a while, it looked like a genuinely new phase.

Elementary chemistry teaches us that water has three phases (or states): solid, liquid, and gas. The Russian finding implied a fourth phase — or at least something distinctly different from the other three phases. You will recall the experimental findings described in the last chapter: water next to common hydrophilic surfaces took on different properties — exclusion zone water was more viscous, more stable, and more ordered than bulk water. While not exactly matching what the Russians claimed to find, these EZ features seem close enough to raise the suspicion that their findings and our findings might overlap.

We begin this chapter by reviewing what the Russians actually found and the international intrigue that surrounded those findings. We will see what useful truths can be extracted from the ensuing debacle. We then focus on the exclusion zone: Is the EZ a simple, organized stack of water molecules, or does it have some other crystal-like organization? And does that structure really constitute a fourth phase of water?

Revisiting the Polywater Debacle

Fig. 4.1 *Boris Derjaguin, pioneering Russian chemist.*

The Russian story began, as mentioned several chapters back, when an obscure scientist named Nikolai Fedyakin discovered that, under certain conditions, water could become unexpectedly stable: it became difficult to freeze and equally difficult to vaporize. It also seemed denser and more viscous than bulk water. Excited by this unusual stability, Fedyakin took his results to the Soviet Union's most prominent physical chemist, Boris Derjaguin (**Fig. 4.1**). Derjaguin was impressed enough to launch a cadre of lieutenants in hot pursuit.

Derjaguin appreciated that capillary tubes were not the only possible materials interfacing with water. Anything touching water creates an interface — from a glass that holds drinking water to the proteins lying inside the cell. All such interfaces create "interfacial" water with properties potentially just as stable as the water lying inside the capillary tubes. Plainly, Derjaguin understood the stakes: unraveling this single phenomenon might hold the key to understanding a good deal of nature. Derjaguin explored the phenomenon meticulously. To assure purity, the water used in his experiments was first allowed to evaporate and subsequently condense inside of scrupulously cleaned glass capillary tubes. It was this seemingly pure water that exhibited such remarkable stability. Yet the very issue of purity ultimately led to Derjaguin's downfall.

Although Derjaguin's work had become well known to the Russian community by the mid-1960s, only later did westerners begin to take notice. Follow-up studies soon began in the US and Great Britain. Soon, everyone became interested in this special kind of water.

Even the press took notice. With its inevitable tendency toward sensationalization, the press aroused concern by fabricating a theory that throwing a thimbleful of this stuff into the ocean might act like a seed crystal, polymerizing all the earth's water supply into one massive blob and rendering it useless for consumption. And so we die.

As a result of this kind of purple cold-war prose, it came as some relief when this polymer-like water, or "polywater," was shown to be an experimental goof. Repeating the experiments, western scientists found that the water contained traces of silica, presumably leached from the walls of the enveloping quartz capillary tube. Hence, the water was impure after all. Although water in a large beaker could hardly be presumed to contain meaningful concentrations of the container material, here the scientists were dealing with extremely narrow tubes with surface-to-volume ratios high enough that the silica concentration inside the water could increase beyond trivial levels; indeed, the concentration was above the detection threshold. Some silica apparently dissolved in the water and, once that contamination was revealed, the Soviets had egg (or silica?) on their faces.

Later, another western scientist took glee in reporting that polywater-like features could be observed when salt was added to pure water — implying that the Russian results might have arisen from summer sweat. Guffaws could be heard reverberating the world around.

Derjaguin himself sealed the coffin of polymeric water by finally conceding that his water had indeed been impure. With this public concession, the world's water supply could be deemed safe after all, rescued from the impending threat of polymeric solidification. Case closed. Debunking polywater had become America's response to the coup of Sputnik. This time, the joke was surely on the Russians.

Although the description of this famous incident has made its way into multiple books, the inside story of the episode has yet to be told. Here are a few relevant tidbits. While traveling in Russia recently, I had the pleasure of chatting with the director of a famous biophysical institute who had been good friends with Derjaguin. They had lived in neighboring flats. The biophysicist told me that the two of them had enjoyed conversations almost daily and assured me that, right up to the time of his death, Derjaguin felt certain that trace contamination was not a decisive issue, despite his published retraction. I heard much the same thing later from another prominent Russian scientist who had been one of Derjaguin's last protégés. Publicly, Derjaguin professed error; privately, he felt certain that he had been on the right track.

Why would a scientist concede a sin he had not committed? The proud Soviet government must have suffered appreciable embarrassment when one of its leading scientists was accused of sloppy scientific technique. The sin seemed to belong to the Soviets. Kept on a short leash by a totalitarian regime, Derjaguin might have been pressured to retract. Retraction would have shifted the blame onto the individual and away from the regime. Blame Derjaguin — not the Soviets.

Political pressure was certainly in evidence on the other side as well. Fearful of Soviet dominance following the Sputnik coup, western scientists were definitely on the defensive. Implicating sweat in the water must have given those westerners a hefty ego boost.

In his book *Polywater,*[1] Felix Franks recounts the events surrounding this famous incident. Although Franks does not question the authenticity of Derjaguin's retraction, one can sense the behind-the-scenes machinations of political forces on both sides that might have influenced the outcome. This political subtext leaves one with an uneasy feeling about what is true and what is not (**Fig. 4.2**).

Fig. 4.2 *The specter of the cold war.*

My own intuition is that both sides were right. Over the years that I have studied water, it has become apparent that obtaining absolutely pure water is next to impossible: no matter the precautions taken, some contamination is unavoidable because water is a universal solvent; it can dissolve practically anything. Hence, Derjaguin's water probably did contain traces of silica and perhaps also traces of salt. The critics might have been on target.

On the other hand, that's where the story becomes interesting. What was questioned in Derjaguin's experiments was only the water — not the correctness of the observations made *using* that water. Let us suppose that Derjaguin's water had been impure, as charged. Then the question reduces to this: in the presence of contaminants, why does water take on the interesting features that it does?

Derjaguin, Fedyakin, and even many western scientists detailed those features in many published papers. Why not consider those features? While I do not advocate careless experimentation, trace contaminants are inevitable and need not automatically disqualify any further exploration. Why throw out the baby with the bathwater?

Keeping these considerations in mind, let us move on to explore the nature of EZ water. EZ water lies near surfaces, just like polywater. Could this similarity be more than coincidental?

Possible Structures of Water Near Surfaces

When we first identified exclusion zone water, many suspected it might be the same as polywater — not in any constructive sense, but in the sense that it might have arisen from a similar experimental error. One prominent physical chemist suggested this possibility to me rather directly, claiming that all he really wanted to do was to save us from the ignominious fate of those involved with polywater.

We responded by adding contaminants to the water. We wanted to see whether contaminants built our exclusion zone in the same way that they had been asserted to build polywater. We found the opposite: practically anything we added to the water diminished the size of the exclusion zone instead of expanding it. The largest exclusion zone correlated with the purest water.

That result told us one of two things: either exclusion zone water was not the same as polywater because it behaved oppositely; or, if EZ water was the same as polywater, then the attacks on polywater might have been unjustly motivated by issues beyond science. At any rate, the polywater specter threw no monkey wrench into our works; we felt justified in exploring exclusion zone water on its own terms.

The name "exclusion zone," by the way, originated with my Australian friend John Watterson, who also suggested the abbreviation "EZ." Now that we know that the exclusion zone does more than just exclude, those monikers might be less than ideal. Nevertheless, EZ does have an easy ring to it, and the term seems to have stuck. For now, we continue to use it.

The central question we faced was the exclusion zone's molecular structure. We felt it had to differ from bulk water, for EZ water was observably more stable, more viscous, and more ordered. But what was that structure?

Stacked Dipolar Water

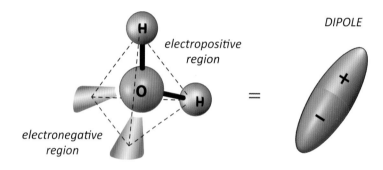

Fig. 4.3 *Textbook structure of the water molecule (left), showing cones of negativity and the region of positivity creating a tetrahedral shape. The separated charges are commonly represented as a simple dipole (right).*

We first considered the most obvious candidate: a simple ordered stack of water molecules. Stacking is possible because the water molecule is a dipole: it comprises an electrochemically negative oxygen atom at one end and two electropositive hydrogen atoms at the other (**Fig. 4.3**). Because of that charge polarization, dipoles naturally tend to stack; hence, it seemed reasonable to consider the EZ's ordered structure in terms of dipole stacking.

Figure 4.4 illustrates this model.

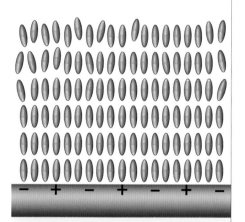

Fig. 4.4 *Stacked dipole model of water ordering. Some loss of order might occur with increasing distance from the surface because of thermally induced motions.*

The stacked dipole configuration would seem the obvious solution to our problem. Beginning at the nucleating surface, water dipoles would stack one upon another, projecting farther and farther from the surface until the disruptive forces of "thermal" (Brownian) motion limit further ordered growth. How far such an ordered crystal might build depends on various assumptions. Most chemists would argue for no more than a few molecular layers, while others argue for an almost unlimited stacking.[2,3]

The stacked dipole model's most fervent advocate has been Gilbert Ling. A world-class scientist, Ling built a comprehensive theory of cell function on the basis of ordered water, implicitly presuming the dipolar arrangement.[4] His cell-function theory has seemed so compelling to me and to some others that, even in relatively recent writings,[5] I could find no reason to question its basis, the simple stacking of water molecules. In fact, that arrangement had seemed the only plausible option.

Later, we found cause to reconsider. Although the stacked dipole model may apply in certain circumstances, newer evidence implied that it could not represent the general case. That evidence, which we will detail later in the chapter, centered on the fact that the exclusion zone bears net electrical charge. Dipoles remain neutral; they cannot build up to yield extensive zones with net charge.

Meanwhile, unaware of the presence of that EZ charge, we proceeded to evaluate alternative candidate structures to make sure we were on a sound course.

Crystalline Water

A good way to deduce possible structures is to start by looking for a precedent. If exclusion arises from ordered water, then a logical approach involves investigating known ordered water structures. Perhaps some variant of one of those structures might suffice.

We saw ice as the most obvious candidate. Ice has a well-known ordered structure. And ice excludes: as it grows, it pushes out molecules and particles, creating a crystal largely free of debris. Could ice's structure offer a clue to EZ structure?

The planes of (standard) ice are arranged in hexagonal units (**Fig. 4.5**). Repeating units create the familiar honeycomb sheet made of oxygen and hydrogen. Protons (*bottom panel*) link each sheet to any sheets lying above and below. Those protons bond oxygen atoms, creating ice's rigid structure. Only every other oxygen is bonded; the remaining oxygens, being electronegative, repel one another, creating the slight pucker evident in each planar sheet.

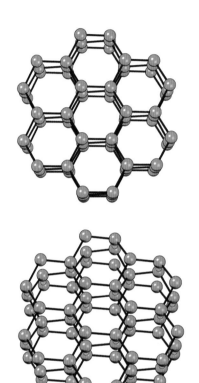

Fig. 4.5 *Structural model of common ice viewed from two different angles. Oxygen atoms are red. Hydrogen atoms (not shown) lie midway along the lines connecting the oxygen atoms. Interplanar protons (blue, bottom) link every other oxygen. The linkages create a pucker, which makes the arrangement of atoms seem less planar and more tetrahedral.*

The exclusion zone, on the other hand, is not rigid; it behaves as a viscous liquid. This means that the structure of ice does not adequately model the EZ's structure. A minor tweak of that ice structure, however, provides a possible EZ candidate. The correct EZ structure requires some fluidity; liquids gain their fluidity when constituent layers can slip past one another. For the exclusion zone, then, a model worth considering is a stack of ice-like planes devoid of those rigidifying interplanar proton linkages. Without the linkages, the planes could slide past one another, conferring the required semi-liquidity. This model seemed promising.

The Charge Issue

Then the charge issue arose. Ice has a neutral net charge. Moving from the ice model to the ice-like model with extracted protons created a problem: the new model required the EZ to have a net negative charge.

Early on, still unaware of the EZ's negativity, we set out to disqualify any model with net charge. After all, the exclusion zone could extend as much as a half-millimeter in width, and expecting to find such a vast zone of charge seemed improbable. The literature on water fell overwhelmingly on the side of its neutral charge, and the familiar dipole model also implied a zero net charge; all of our scientific experience left us confident that an uncharged region was more likely than a charged region. So we anticipated easily disproving the ice-like model or any model bearing net charge.

To accomplish that disproof, we designed a straightforward experiment (**Fig. 4.6**, *opposite page*). The lab's experience measuring electrical potentials in living cells made it easy to set up a comparable experiment in and around gels. We used microelectrodes. As their name implies, microelectrodes taper to extremely fine tips, making possible micron-scale spatial resolution. We planted one microelectrode remotely as a reference. A motor positioned another microelectrode progressively closer to the gel surface in order to chart the near-surface electrical potential. We could thus determine whether the exclusion zone was charged.

To our surprise, we found that the EZ was indeed charged — negatively charged.

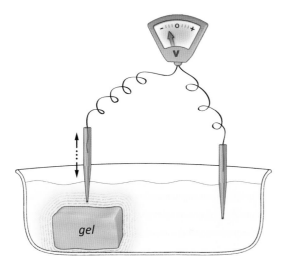

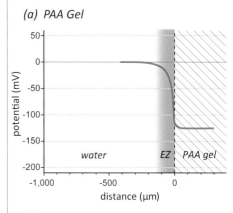

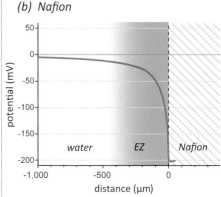

Fig. 4.6 *Experimental setup used for measuring the electrical properties of the exclusion zone. Reference electrode at right.*

(a) PAA Gel

(b) Nafion

Fig. 4.7 *Electrical potentials measured near a polyacrylic acid gel (a), and a Nafion sheet (b). The zone of negativity correlates with the width of the EZ, which differs in the two cases.*

Figure 4.7a shows a representative result. With the motor-driven electrode positioned initially well beyond the exclusion zone and therefore in bulk water, we measured a potential difference of zero. That was expected. As the electrode approached the interface, it began to report negative potentials, the magnitude increasing with proximity to the surface. Immediately outside the gel, the 120 mV negative potential remained steady even as the electrode continued to advance inside the gel. In the next experiment we removed the gel and replaced it with a sheet of Nafion. With Nafion, the magnitude of the near-surface negative potential rose as high as 200 mV (**Fig. 4.7b**).

For both specimens, the region of negative potentials extended rather far from the interface: to approximately 200 μm for the gel and more than 500 μm for the Nafion. It appeared that the EZs bore negative charges.

This result was quite unlike the neutrality we had anticipated. Rather than quickly eliminating the ice-like model, the negative charge lent it support. At the same time, it disqualified the dipole model; dipoles contain no net charge.

It seemed we were making progress. However, my colleagues were not shy to suggest a possible oversight. Having trained as an electrical engineer, I should have been savvy enough to figure it out; instead, my students had to remind me that negative electrical potentials could arise from a negative surface charge: if the material surface were charged, then the effect of that charge might be felt at some distance. Hence, the negative potential inside the EZ needn't necessarily implicate any net charge. Ouch!

It took me a few minutes to recall that matters were different in water. Surface charges cannot extend their influence very far in water, for counter-ions in the water would inevitably gather and mask the material's surface charge; beyond a short distance from the surface, you'd measure zero. This consideration does not depend on the double-layer theory; fixed charges will *always* attract available opposite charges in a liquid. So, my colleagues' suggestion might apply in a vacuum but not in liquids like water, where mobile charges will cancel any long-range influence.

Nevertheless, we checked for the presence of negative charge inside the EZ. One test sought to determine whether we could find a corresponding pool of positive charge elsewhere. The EZ builds from neutral water. Starting with a neutral entity and ending with a negatively charged entity makes no sense; surely if the EZ contains negative charge, then a commensurate pool of positive charge must lurk elsewhere.

Fig. 4.8 *Measuring the pH of water next to an immersed gel. The gel occupies an appreciable fraction of the beaker's volume.*

That positive charge ought to appear in the form of protons, because protons are water's only positive charge carriers. If the EZ bears net negative charge, then we should find a zone replete with protons — i.e., a zone of low pH.

As detailed in Chapter 5, we tested for that low pH by inserting a bulky gel into a beaker of water (**Fig. 4.8**). Exclusion zones quickly built next to the gel. A pH probe inserted into the water beyond the EZ showed that pH values dropped substantially, sometimes to values as low as 2, or even 1. The magnitude of the drop was astonishing. Those ultralow pH values indicated that the water beyond the exclusion zone contained protons in huge concentration.

Finding those protons confirmed the zone of positive charge that was anticipated; that positive charge complements the EZ's negative charge. Overall, the body of water seemed as neutral as the water initially used for building the EZ. Apparently, as the EZ builds, water's charges separate into negative and positive components. We had managed to identify both components.

This news seemed both bad and good. It was bad because it confirmed that Ling's stacked dipole model, which I had endorsed in my 2001 book, was inadequate; dipole models contain no net charge. I had evidently erred. On the other hand, it was good news because the ice-like model seemed promising: it could account for the EZ's negative charge as well as its semiliquid nature. Best of all, the model had precedent — it was not magically pulled from a hat.

Would it surprise you to learn that somebody else had proposed the very same model decades ago? In a 1969 lead article in the respected journal *Science*, ER Lippincott and collaborating chemists from the University of Maryland hypothesized practically the same structure — for polywater. Their original figure is reproduced in **Figure 4.9**. That polywater model made no headway at all. Because of polywater's discrediting only several months after the model's publication, nobody took the trouble to consider the proposed structure in any depth. Any and all attempts to investigate the nature of polywater quickly got laid to rest.

Now the polywater model gains fresh relevance. We want to know more about it. What had scientists learned before negative publicity brought a sudden stop to that field's progress?

Polywater Revisited

Similar to the ice-like structure we have been considering, the polywater model envisions a stack of honeycomb sheets made up of hydrogen and oxygen atoms. The authors of the fateful *Science* article had deduced that structure from a broad range of amassed physicochemical data. The data included Raman spectra, extreme freezing and boiling points, and high density. The structure depicted in **Figure 4.9** fit the experimental data best.

Several aspects of that *Science* article struck a resonant chord. First, the article reminded me that nature likes hexagonal structures. You see them throughout the domain of organic chemistry. You also see this structure in graphite, where constituent honeycomb (graphene) sheets can slide easily past one another, resulting in low friction. Being common, hexagonal sheets seem a natural option to consider.

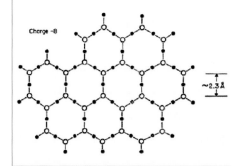

Fig. 4.9 *Molecular structure of polywater, proposed by Lippincott and colleagues.[6] Oxygen molecules denoted by open circles, hydrogens solid. Sheets stack to yield volumetric structure.*

Second, the authors provocatively state that the substance in question is *not* water. The substance is certainly built of oxygen and hydrogen, but their arrangement in a hexagonal lattice bears little resemblance to their arrangement in the water molecule. This new substance, they asserted, "should not be considered to be or even called water, any more than the properties of the polymer polyethylene can be directly correlated to the properties of the gas ethylene." The authors considered it clear that this entity was chemically distinct from water.

A third feature, which really made me stand up and take notice, was the ratio of hydrogen atoms to oxygen atoms. As everyone knows, these atoms have a 2:1 ratio in bulk water. In this planar structure, their ratio is 3:2. This feature may not be immediately obvious, but **Figure 4.10** presents a simplified method for verifying the ratio.

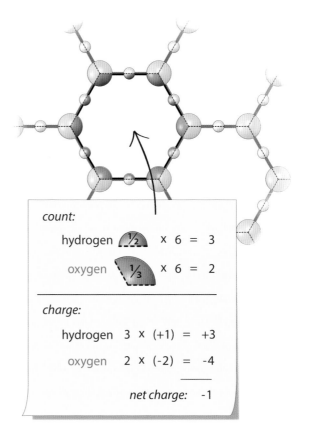

Fig. 4.10 *Computation of the net charge of each hexagonal unit. To make the count, represent each atom as a sliceable pie; then count all the pie fractions lying within a given hexagon, taking care to remember that the oxygen pie's charge is minus two while the hydrogen's is plus one. The resulting ratio of hydrogen to oxygen is 3:2, and the net charge is of the hexagon is -1.*

This numerology matters because the familiar 2:1 ratio confers neutrality. Two electropositive hydrogen atoms balance one electronegative oxygen atom, so the water molecule is neutral. The lattice, on the other hand, has an unbalanced ratio — yielding one negative charge per unit hexagon.

The authors made special note of this feature on the upper left corner of their figure (see **Fig. 4.9**), but they paid scant attention to its potential significance. In a matter-of-fact way, they presumed that positive charges lodged in between the negatively charged planes would neutralize most of the charge. The model's essential point was that the planes themselves were negatively charged.

The Lippincott polywater model is essentially the same as the model put forth in these pages. The polywater model derived from strict physicochemical reasoning, while our model derives mainly from precedent and logical inference. Both routes lead to essentially the same result: a honeycomb sheet with a hydrogen-to-oxygen ratio of 3:2.

This 3:2 ratio has been experimentally implicated. A report in a prominent physics journal created a stir with a finding of this very ratio: when protons and neutrons bounced off water molecules, the scattering pattern implied $H_{1.5}O$, not H_2O,[7, w1] Of course, the ratio 1.5:1 is the same as 3:2.

The key feature for both of these models is the hexagonal arrangement of atoms, raising the question whether hexameric (i.e., hexagonal) figures are experimentally observable. The answer is yes. Researchers have identified water hexamers next to diverse surfaces, including metals,[8] protein subunits,[9] graphene,[10] and quartz.[11] Near-surface hexamers have also been found in supercooled water.[12] And water adsorbed on mica showed a preponderance of 120° angles, which was interpreted as evidence for hexagonality.[13] The water next to many surfaces is evidently hexagonal, which agrees with the proposed model.

Of the evidence showing hexagonality, a study worthy of comment is the one investigating water droplets encapsulated by protein.[14] The specific protein is subunit-c of ATP synthase, an ancient protein that appears throughout phylogeny. In dry conditions this protein forms an encapsulating shell around water, preventing evaporation.

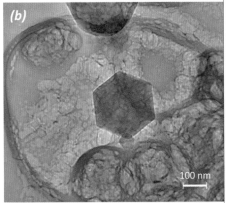

Fig. 4.11 *Protein-encapsulated water.[14] Encapsulation produces spheres (a) seen in the scanning electron microscope, and geometric figures (b) seen in the transmission electron microscope. Diffraction pattern (c) obtained from geometric figure shows hexagonal order.*

Fig. 4.11 shows two examples of these shelled structures: spherical capsules (*panel a*) and geometric capsules (*panel b*). Diffraction patterns obtained from geometric capsules show that the enclosed water has hexagonal order (*panel c*). Further, the hexagonal unit spacing, 0.37 nm, is close to that shown in **Figure 4.9**. Thus, hexagonal order can be seen in substantial volumes of water near surfaces.

Another anticipated feature of these models is the absorption of ultraviolet light. Absorption at or around the 270 nm (UV) wavelength is expected when electrons are "delocalized," i.e., free to move through the structure. That situation exists most commonly in aromatic (ring) structures, and also in so-called "crown-ethers," whose oxygen-containing hexagonal structures are similar to the structure under consideration. Thus, the confirmed EZ absorption at 270 nm (**Fig. 3.13**) adds evidence in support of a hexagonal structure.

So anticipated UV absorption is confirmed, hexamers are experimentally detectable, and two independent sets of considerations lead to essentially the same hexameric model. This supportive evidence provides impetus to consider the model more seriously. Let us press on.

Stacking Honeycomb Sheets

In exploring the proposed model's explanatory power, we must first determine how the honeycomb sheets stack to form the exclusion zone; after all, the EZ is a three-dimensional entity, not a single sheet. We also need to understand how the initial EZ layer forms. Let us first deal with how the sheets stack.

The simplest stacking model puts the hexagons of all planes in register. You could look down the stack of hexagons and see all the way through.

This in-register arrangement is attractively simple — but impossible. To appreciate why, look back at the top panel of **Figure 4.5**, which shows an example of planes in register. Suppose you remove the interplanar protons (evident in the bottom panel). That creates the planar stack under consideration. Removing the proton "glue" juxtaposes negatively charged oxygen atoms in one plane with negatively charged oxygen atoms in the next plane. That would create interplanar repulsions in huge numbers. The structure would immediately fly apart.

A more natural way for the planes to hold together is by shifting them out of register (**Fig. 4.12**). If the negatives of one plane lie opposite the positives of the next plane, then those planes could stick by electrostatic attraction.

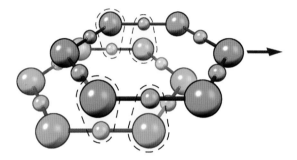

Fig. 4.12 *Shifting one plane relative to another puts opposite charges next to one another, creating attractions.*

Such a planar shift is theoretically realizable in two ways, but only one can do the job (**Fig. 4.13**). The first mode involves a shift in the direction perpendicular to a hexagonal strut (***panel a***); the second involves shifting along a strut (***panel b***). In the first model, no degree of shift leads to any regular apposition of opposite charges, and hence no appreciable stickiness. In the second model, shifting by half the oxygen-to-oxygen spacing brings many opposite charges into apposition: one third of all planar charges stick. This abundant stickiness confers ample cohesion. The cohesion in turn confers high density (see Chapter 3, *p. 38*). Thus, this second model seems to work.

(a) *(b)*

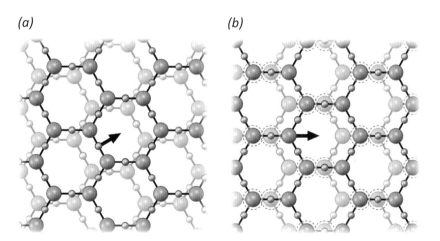

Fig. 4.13 *Possible plane-stacking arrangments involving linear shifts. Only the shift shown in the right panel yields a stable structure with overlapping opposite charges.*

The planar shift also creates some repulsions: nearby like-charged atoms from respective planes repel one another. However, repulsions are fewer than the attractions; and those repulsive forces push away the repelling atoms, thereby weakening the net repulsive force. Indeed, our computations have shown that the attractive forces easily win out.

So the second model yields a stable structure that sticks together naturally. This model yields predictable mechanical behavior: semi-solid when left alone, yet able to flow in response to an imposed shear force. Its behavior should resemble gelatinous egg white.

Variants on that simple stacking theme lead to interesting structural variants. In **Figure 4.13b**, successive planes of the stack are shifted rightward; but they could just as easily be shifted leftward. With these two options, you could build a left-tilted or right-tilted edifice. These variants could perhaps explain the mirror-image constructs mentioned in Chapter 2.

In fact, the shift direction need not be restricted to left or right alone; planar shift could occur in any one of the six strut directions, leading to endless stacking options. We can even realize helical stacking (**Fig. 4.14**): Start with a base plane, shift the plane above it in the

Fig. 4.14 *Shifting successive planes by 60° yields a helical structure.*

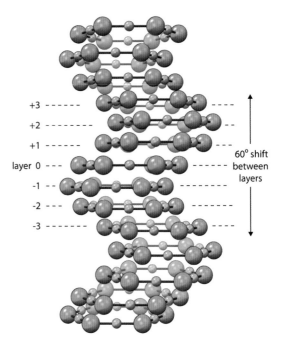

direction along a strut, shift the next plane 60° to that strut, the next an additional 60°, *etc*. The helical pitch would then comprise six planes. Larger pitches are theoretically realizable, even irregular pitches. This helical feature may be especially relevant for biology, where EZ water needs to interface with the helically wound proteins and nucleic acids.

In sum, pouring water onto a hydrophilic surface triggers EZ growth. Water is the raw material. From this raw material, EZ honeycomb layers build. Those EZ layers can slide past one another if sufficient shearing force is applied; but ordinarily the planes stick to one another, creating what is seen macroscopically as the EZ (**Fig. 4.15**).

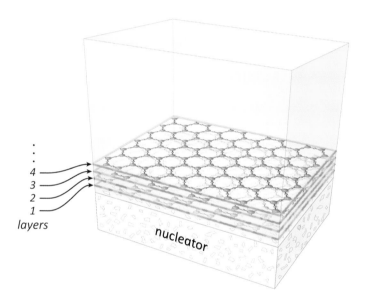

Fig. 4.15 *Buildup of honeycomb planes from bulk water. Hydrophilic surface nucleates EZ growth, which progresses layer by layer.*

The Initial Layer

How does the EZ construction process begin? Hydrophilic surfaces generally contain oxygen atoms, and one possibility is that those surface-oxygen atoms form a template. If enough of those atoms' positions correspond to the positions of oxygen atoms on the EZ honeycomb, then the surface itself could be thought of as the first EZ plane. Additional planes would then easily stack from that template plane.

Of course, no material surface provides a perfect match. Surfaces differ as to atomic arrangements and may have different negatively charged atoms instead of oxygen. Some surfaces might therefore prove less adept at nucleating EZ layers; they would be considered less hydrophilic.

A subtle implication is that the nucleator imparts information to the EZ layers. A patch of missing oxygens on the template, for example, implies a corresponding miss on the first EZ layer, and so on; the missing patch could project through many layers. If so, then the EZ would contain information about the nature of the nucleating surface. To the extent that the EZ is stable over time, that information could be retained.

Another implication is that nucleating EZ growth needs only the template, i.e., a single molecular layer. That explains why EZs can grow even from a monolayer (**Fig. 3.9**).

Materials lacking surface charges, on the other hand, should have scant capacity to template EZ growth, as should surfaces whose charges utterly fail to match the standard honeycomb pattern. All such surfaces would be classified as hydrophobic — water hating, or fearing. Only charged hydrophilic surfaces can nucleate EZ growth.

Even with suitable templating charges, a relevant factor for EZ growth may be surface roughness. Slight roughness is not an issue: if a surface is only slightly rough on a molecular scale, then the initial EZ layer should conform to surface bumps, pits, and ridges; the planar stack will adopt a slightly wavy configuration. More serious roughness could introduce discontinuities: instead of one extensive EZ planar stack, numerous mini-stacks would grow from the plane of each surface tilt. If those stacks sterically interfere, then EZ growth will be impaired. Such a template might not generate EZs as extensive as those nucleated by flatter surfaces. Some preliminary evidence from our laboratory supports that implication.

Hence, the template itself is not the sole agent governing EZ size. Templates nucleate EZ growth by providing suitable atomic matches. Strongly hydrophilic templates provide better matches and therefore nucleate more robust EZs; however, surface roughness and other significant factors (see below) influence ultimate EZ size. The template is only one among several size determinants.

Lattice Erosion and EZ Size

So far, so good — but here's a problem. Identical EZ planes should produce identical electrical potentials. In fact, the EZ's electrical potential falls off with distance from the nucleating surface (**Fig. 4.7**). The planes cannot be identical. To account for this falloff, planar charge must diminish with distance from the nucleating surface. This diminution can occur in either of two ways: by removing negative charge or by adding positive charge. Both are possible.

Removing negative charge from the lattice means eliminating oxygen atoms; the more oxygen atoms removed, the less the plane's overall negativity. **Fig, 4.16** shows that limited removal is structurally tolerable: so long as the number of removed oxygen atoms is not excessive, the planar lattice will not disintegrate. Even the loss of every other oxygen need not compromise the lattice, because interplanar attractions can promote lattice stability. If oxygen-removal "defects" increased with distance from the nucleating surface, then more distant planes would be progressively less negative.

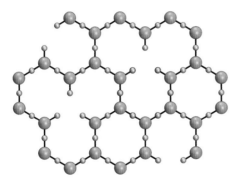

Fig. 4.16 *Diminishing planar charge. The example shows how removal of oxygen atoms from the hexagonal lattice can occur without impairing structural integrity.*

One way to realize this oxygen loss involves lattice erosion. Lattice negativity motivates positively charged protons to penetrate back into the EZ. Actually, it's not protons *per se* that are motivated, because the protons are short-lived; the protons immediately latch onto water molecules to form hydronium ions. Ordinarily, those hydronium ions cannot enter the EZ lattice; lattice tightness prevents penetration.

However, openings such as those schematized in **Figure 4.16** create opportune sites for invasion, as would lattice irregularities created by surface roughness. Were the invading hydronium ions to combine with proximate oxygen atoms in the lattice, the combination would create water. This would erode the lattice. The most serious erosion should occur where hydronium ions begin penetrating, assuring the greatest compromise of negativity at the EZ's distal plane — matching what we observe experimentally.

The invading positive charges may also stick in the space between planes. Especially if a proton were to break free of a hydronium ion, it could bridge closely spaced oxygen atoms of successive planes (see **Figure 4.13b**). Again, this would most likely occur where protons are more abundant, near the distal plane of the lattice. By adding positive charge, those protons produce the same result as oxygen erosion: diminishing negativity in the planes farther from the nucleating surface.

The extent of lattice erosion could influence exclusion zone size. Extremely hydrophilic surfaces with relatively few lattice defects should build EZs suffering limited erosion. With less hydrophilic surfaces having more defects, positive ions could enter more easily, eroding the lattice and compromising EZ size. This may explain why less hydrophilic materials generate smaller exclusion zones.

The presence of lattice defects is reminiscent of semiconductors, where lattice defects in the crystalline material lead to structures with excess electrons or excess "holes," referred to as n-type or p-type semiconductors, respectively. The structure of the EZ is more like the n-type, with excess electrons borne by oxygen atoms. Thus, we anticipate some semiconductor-like features for the EZ, and later, we will see that such features are present. For now, suffice it to say that the lattice defects provide a mechanism for governing EZ size through lattice erosion.

Positively Charged Exclusion Zones?

The attentive reader may have spotted a curious anomaly in the section above. Plucking a few oxygen atoms from the lattice diminishes negativity: the more oxygens you eliminate, the less negative the lattice becomes. This leads to the obvious question of what happens in the extreme. Plucking every other oxygen would shift the net charge beyond zero, all the way into the positive numbers. You can go through the exercise yourself and verify that the net charge per hexagonal unit of the standard lattice shifts from -1 to +1.

At first blush, this feature seems curious, as it implies the possible existence of positively charged exclusion zones; so far, we have considered only negative EZs. Yet, if the proposed model for the EZ is adequate, we might expect to find positively charged exclusion zones as well; their structural framework would remain the same except that they would have far fewer oxygen atoms.

Such positive exclusion zones exist, although they are less common than the negative ones. We found them next to certain polymers and metals.[15] Ion-exchange gel beads provide an example. Commonly used for physical separations, these half-millimeter beads come in two varieties: anionic and cationic. Both types nucleate exclusion zones (**Fig. 4.17**), but EZs next to the cationic beads bear a positive net charge.

Figure 4.18 shows evidence for that positive charge. The figure demonstrates that the spatial distributions of electrical potential next to the cationic and anionic beads practically mirror one another. One is the standard negative type, whereas the other shows a correspondingly positive potential. Regions outside the positively charged EZs show a higher pH, instead of the usual lower pH seen beyond the negative EZs.[15]

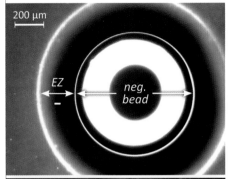

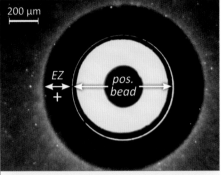

Fig. 4.17 *EZs surround both negatively and positively charged beads.*

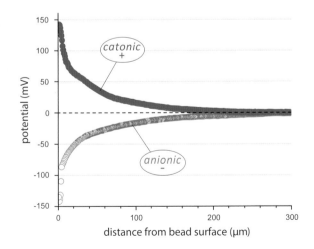

Fig. 4.18 *Electrical potentials recorded next to cationic and anionic beads.*

Thus, both potential distributions are realizable. Features of the positively charged exclusion zones seem more or less the inverse of the negatively charged ones.

If the positive EZs contain many fewer oxygens, then you might expect those lattices to be more fragile, since defects weaken the lattice. We have confirmed that fragility. Positive EZs are quirky excluders that can be broken up relatively easily with minor mechanical perturbation. As a result, my colleagues hesitate to study them in the laboratory. Nevertheless, those positively charged exclusion zones do exist.

The proposed structural model evidently has enough versatility to accommodate both types of exclusion zones, negative and positive; separate structural models are unnecessary. This attribute appeals because we anticipate that nature favors simplicity. Furthermore, if EZs occur as commonly as we will later show, then the positive EZs' fragility may explain their relative rarity; they don't often survive.

A Fourth Phase of Water (And Why Some Chemists Suffer Apoplectic Fits)

A significant feature of the proposed EZ model is its similarity to ice. This similarity should not surprise: on that very basis, we derived our model of the EZ's structure. On the other hand, if ice qualifies as a phase of water, then the EZ might likewise qualify — perhaps as the long sought "fourth phase" of water suggested a century ago by the prominent physical chemist Sir William Hardy.

For the EZ to qualify as a phase (sometimes referred to as a "state"), it would need to satisfy certain criteria: it should be unique and spatially bounded, and it must exist in a significant quantity. These criteria seem satisfied for water's three classical states (although the coming chapters raise some question about vapor). They are also true for the EZ: exclusion zones are bounded, uniquely structured, and can project from a surface by distances up to a full meter (see **Fig. 3.8**). The EZ seems as qualified as ice for consideration as a phase.

On the other hand, referring to the EZ's extensiveness can occasionally throw otherwise sober chemists into fits. How could any structure built of water molecules extend millions of layers from a nucleating surface? Educated to think that the disruptive effects of thermal motion will limit ordering to a few molecular layers, some chemists are prone to view the concept of long-range order as a nonstarter; it simply cannot happen.

However, we are not proposing a structure composed of stacked water dipoles, but of stacked planes. The two differ. Chemists might view a stack of dipoles as similar to a stack of bricks, made wobbly by the disruptive effects of thermal motion (**Fig. 4.19,** *left*); since these disruptive effects are additive, the stack cannot grow very high before faltering. It bears repeating that we are not positing a stack of dipoles but a stack of planes (**Fig. 4.19,** *right*). Each plane is extensive, and the more extensive the structure, the lower the thermal agitation. So any disruptive effects should be far less pronounced in the planar stack than in the dipole stack. One might hope, therefore, that this planar model would prove less likely to provoke chemists into reflexive fits of outrage.

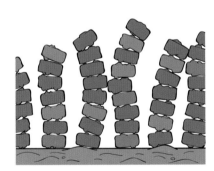 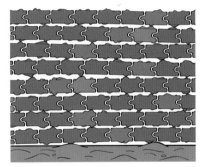

Fig. 4.19 *Stacked dipole model may lead to wobbling and disorder (left); however, disorder is minimized when elements interconnect to form extensive planar structures (right).*

In a different vein, the planar model helps us to reconcile an issue that chemists have not resolved: why gels retain so much water. Gels hold their water. Remember that common gels don't leak, even when their fractional water content exceeds 99.9 percent of their total mass (**Fig. 1.1**). Now we can venture an explanation of that phenomenon. The gel matrix comprises numerous hydrophilic strands. The strands' surfaces convert bulk water into EZ water. The EZ planes stick to those nucleating strands and also to one another; hence, your gelatin dessert remains hydrated. The EZ water doesn't dribble out.

Finally, the proposed structure makes clear why exclusion zones exclude. They exclude because it is only through the hexagonal openings that solutes can enter the EZ lattice and those openings are small. The actual impediment is even more formidable: because successive EZ planes are out of register, the effective openings are narrower than the planar hexagonal openings (**Fig. 4.15**). The lattice is extremely tight, and therefore highly exclusive of solutes. Only protons and smaller entities are small enough to penetrate.

On the other hand, protons ordinarily don't exist as distinct entities; they stick to water molecules to form hydronium ions, which are far bulkier than protons and therefore excluded. Later (Chapter 17), we will see how protons freed from those parent water molecules can penetrate into the EZ lattice to form ice.

Apart from those freed protons, it appears that *all solutes get excluded* — at least from lattice regions devoid of openings. Exclusion of even hydronium ions, with their positive charge, ensures maintenance of the electrical potential difference between the EZ and the water beyond. That's why we can measure the same potential difference over long periods of time.

• • •

The sustained charge separation between the EZ and the bulk water zone beyond has repercussions for much of what follows. That charge separation constitutes a "battery." The character of the battery and the nature of the energy that keeps the battery charged will prove pivotal for understanding practically all phenomena involving water.

Summary

In formulating a structural model of the exclusion zone, we first considered stacked dipoles. While simple, logical, and historically grounded, dipoles remain stubbornly neutral; they cannot account for the exclusion zone's net charge. Therefore, the dipole model failed. We found the honeycomb sheet model more promising, its hexamers lying out of register with those of adjacent sheets. This model could account for the EZ's net charge; plus, it has the advantage of precedent because of its similarity to ice.

In this stacked sheet model, local charge depends on the density of electronegative oxygen atoms. Thus, local electrical potential can range from extremely negative values to zero and all the way into the positive values characteristic of some exclusion zones. The basic structural framework has enough versatility to account for all types of exclusion zone.

Real EZs differ from generic EZs. Generic EZs contain full hexagonal lattices. Real EZs are less regular: they may lack oxygen atoms and hydrogen atoms at positions that reflect the nucleating surface's charge distribution, and they may suffer erosion.

Exclusion zones seem both extensive enough and distinct enough to qualify as a separate phase of water. Recognition of this "fourth phase" has only just begun. Its elucidation promises to shine light on what transpires when water touches practically anything in sight.

5 Batteries Made From Water

The skies above unleash a flash of lightning, discharging hundreds of thousands of volts of raw energy to the earth's surface. Those lightning strikes occur so frequently around the world that, according to atmospheric scientists, the earth's surface cannot dissipate the accumulating negative charge, leaving it electrically negative. Standing on the ground, your nose is about 200 volts more positive than your toes.[1]

While lightning and its electrical consequences are not the subjects of this chapter, charge definitely is. Like clouds, EZs contain concentrated electrical charges. Those charges carry potential energy, just like the thundercloud's charges. And the consequences can be equally impressive.

Consider biology. Charged entities such as membranes, proteins, and DNA all interface with water; exclusion zones should appear in abundance. Those EZs bear charge, which means they carry electrical potential energy. Since nature rarely discards available potential energy, EZ charge may be used to drive diverse cellular processes ranging from chemical reactions all the way to fluid flows. Opportunities abound.

On the other hand, isolated bodies of charge are rare. Charged bodies usually sit apposed to bodies of opposite charge — e.g., on the opposite faces of a biological membrane. We therefore anticipated a similar pairing of the concentrated charge in the EZ with a balanced accumulation of opposite charge lying beyond the EZ

So let us pick up and expand on our inquiry into whether the EZ might have a companion pole of opposite charge (Chapter 4), exploring some consequences of any such pairing.

Charges Beyond the Exclusion Zone

In considering whether the EZ system contains two poles instead of one, we need to keep in mind that the exclusion zone builds from plain old water, which is neutral. If a neutral substance spawns a charged EZ, then an equal and opposite charge must lie somewhere else; otherwise, the law of conservation of charge would be violated. Violating laws cannot lead us anywhere useful.

We expected, therefore, to find a zone of opposite charge just beyond the EZ. As the negative charge builds within the EZ, a corresponding zone of positive charge should build just beyond. That region would then acquire many protons. Since a high proton concentration means low pH, we suspected early on that a low pH zone might exist in the water exterior to the EZ.

To test this notion, we put a gel into a beaker of water and positioned a pH probe just outside the gel's EZ (see **Fig. 4.8**). We would have been excited to see the pH drop by one unit, indicating a tenfold increase in proton concentration; but, (as mentioned in Chapter 4), we got a more dramatic result. Next to the EZ of the polyacrylic acid gel, we often found pH drops of three or four units and occasionally more. Since the pH scale is logarithmic, this meant a ten-thousand-fold increase of proton concentration. That result amazed us.

We found that altering the experimental setup could change the magnitude of that pH drop. For example, we varied the size of the beaker relative to the gel; with a beaker very much larger than the gel, we saw only a modest pH drop; but, using a beaker barely larger than the gel so that the protons had practically nowhere to go, we obtained more impressive pH drops.

Little in classical chemistry helped us to understand what we observed. The results were so dramatic that our more conservative laboratory members became uneasy. One bright lad, well versed in classical chemistry, simply could not believe the result and moved on to another project. I wasn't absolutely convinced myself at first.

We had a lingering concern that perhaps those protons did not really accumulate as a consequence of EZ buildup. Their surprising accumulation would lose significance if we found that they merely leaked from the gel. We soon figured out how to test for this alternative: if the protons came from the gel, then their accumulation could not exceed some fixed value; after all, no gel can supply an infinite number of protons. We found that pre-immersing the gel into a succession of water baths in order to draw out any and all releasable protons made little difference: in subsequent trials, we saw similar drops of pH. It seemed that the proton buildup really did arise as a result of EZ buildup, as we had suspected.

That conclusion was satisfying — like a cool drink on a hot summer's day. It also reassured us because evidence of high positivity lent credence to the evidence of the EZ's correspondingly high negativity. To some physicists, this quasi-stable concentration of negative charge had seemed beyond common experience and therefore difficult to accept. Finding opposite charges elsewhere reassured us that we were on a productive track.

Proton Buildup

To nail down the dynamics of the proton buildup, we used a miniature pH probe. The probe was small enough that we could track the local pH change at a series of distances from the EZ-nucleating sample (**Fig. 5.1**, *top*). For the sample, we used a sheet of Nafion secured to the bottom of a chamber. We filled the chamber with water and then observed the proton buildup.

The bottom panel of **Figure 5.1** shows records of pH changes detected at several distances from the sample. At a distance of 1 mm, the pH began to drop within a few seconds, reaching a low point in 15 seconds and then recovering partially as protons spread to more distant regions. At a 5 mm distance, the pH change started later; at 10 mm, it began still later. Eventually, the pH came to roughly the same final value, lower than the initial pH value, at each point of measurement.

Figure 5.1 shows pH values changing later for measurements taken farther from the Nafion surface. These successive delays imply a wave of protons originating from near the sample and diffusing away. The wave seems likely to originate at the EZ's outer

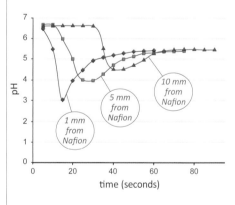

Fig. 5.1 *Time course of pH change following the addition of water to a sheet of Nafion. We measured the pH at three points, as indicated. The pattern suggests a progressively rising wave of protons.*

edge, since that's where the EZ builds (**Fig. 5.2**). We can anticipate equilibration over time: protons repel one another and will eventually distribute more or less uniformly — at least those protons situated well beyond the grip of the EZ's negativity. The time of equilibration is not fixed; it should depend on the physical nature of the system and particularly on how much space the protons have for spreading out.

Proton Distribution: pH-Sensitive Dyes

Because the proton issue seemed critical, we employed an additional proton detection tool: pH-sensitive dyes. These dyes, like the ones used in litmus paper, change color depending on pH.

Figure 5.3 shows a representative example of the color distribution observed beyond the exclusion zone. The red-orange color near the EZ, according to the color calibration chart, indicates a pH value of three or below — i.e., numerous protons. At greater distances from the EZ, the pH was also lowered, but less dramatically. Hence, the pH dye technique largely confirmed what the pH probe showed: abundant protons in the water just beyond the EZ. Another reassuring result.

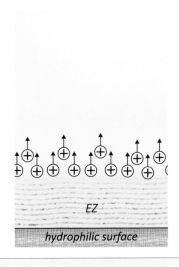

Fig. 5.2 *Protons generated at the leading edge of growing exclusion zone.*

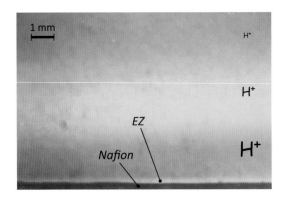

We tried to take the dye results a step further and use them to compute the number of protons, but we only managed to arrive at crude estimates. We were stymied by our inability to compute the number of protons abutting the EZ: the exclusion zone's negative charge should draw many positive protons, but the dye's measurement range could not handle such high concentrations.

Fig. 5.3 *Proton distribution beyond the exclusion zone (viewing the wide face of a narrow chamber). This image, obtained shortly after pouring water with pH-sensitive dye solution into the chamber, shows that the EZ excludes the dye. The pH value immediately beyond the EZ is 3 or below (red-orange), indicating numerous protons. The expectedly higher proton concentration abutting the EZ lies beyond the dye's detection range.*

Undaunted, we set up a chamber like the one shown in **Figure 5.3** to measure what we could. We found an estimated 10^{15} to 10^{16} protons distributed throughout the space beyond the EZ. For comparison, we estimated the number of EZ electrons. This was computed from the lattice structure and the measured potential distribution. That number, 10^{18} to 10^{19}, was substantially higher than the measured proton count. Two uncertainties might account for the difference: the unmeasured protons abutting the EZ; and/or the reduced negative EZ charge created by less-than-full oxygen occupancy. Hence, we could not definitively answer the question of the exact match between positive and negative charge.

Later, we tracked the released protons with yet another method: a setup that continuously refreshed the near-EZ water (**Fig. 5.4**). We used a hollow Nafion tube. The tube's inner surface nucleates a ring-like exclusion zone just inside, which drives protons into the core (*a*). We refreshed those core protons by continuously infusing fresh water through the tube (*b*). Since the annular EZ tends to cling to the tube material, much of the tube flow occurs in the core. We found that the exiting water had a lower pH value than the entering water; the pH difference exceeded one unit and never diminished — even after 30 minutes of continuous flow.[2] While we still couldn't resolve the

(a)

Nafion tube filled with water

EZ

protons

(b)

fresh water

Nafion tube

protonated water

Fig. 5.4 (a) *Annular EZ releases protons to core.* (b) *Passing fresh water through the tube sweeps out the released protons.*

quantitative issue, we did establish that the passing water continued to receive protons from the annular EZ without diminution, even over extended periods of time.

We even found evidence of released protons in suspensions of microspheres. As hydrophilic entities, those spheres ought to envelop themselves with exclusion zones, presumably in the form of shells. Any such shell-like zones might be small enough to evade detection by microscope; however, we should be able to measure the corresponding pH change in the water. The larger the number of microspheres, the larger should be the pH change. **Figure 5.5** confirms that expectation.

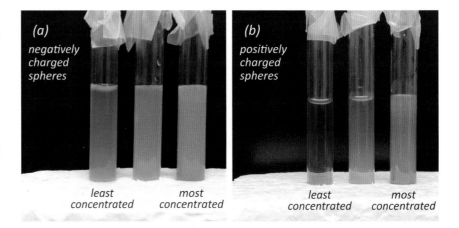

Fig. 5.5 *Addition of microspheres alters water's pH. (a) Carboxylate microspheres, 1 μm diameter. Increasing microsphere concentration changes dye color toward red, indicating lower pH. (b) Positively charged amino microspheres change dye color toward green, indicating higher pH.*

The collection of results described above confirms the measurements made using the pH meter: positive charges in the bulk water consistently appeared next to the EZ's negative charges. Water molecules have effectively split into negative and positive, creating something that looks suspiciously like a battery — a chemical factory with separated charges.

Battery-like charge separation remains evident even when the exclusion zone grows into idiosyncratic configurations. **Figure 5.6** shows an example. This image came from the same experimental setup as did **Figure 5.3** but was obtained at a later time. (The "bumps" in the Nafion arise as hydration induces buckling; they do not affect the result.) By the time we obtained this image, the EZ had begun sprouting pole-like projections similar to those of **Figure 3.7**. You can see red surrounding each projection. The red represents the density of protons

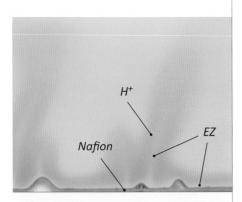

Fig. 5.6 *Similar to Figure 5.3, taken at a lower magnification and later time. Note zones of low pH (red) surrounding each vertical EZ projection.*

Cellular Batteries: Nerves, Pain, and Anesthesia

Ouch! The stove is unexpectedly hot. You reflexively withdraw your hand in order to avoid the unpleasant consequences that would otherwise follow.

Nerves mediate that withdrawal: your nerves signal your brain to quickly pull your hand. That signaling mechanism is electrically based: nerve cells bear negative charge, while the regions outside those cells are positively charged. Noxious stimuli trigger a local discharge that propagates along the nerve to your brain. Thus, charge separation is a central feature of nerve-signal transmission; each nerve behaves like a dischargeable battery.

How does this separation of charge occur? According to the prevailing view, the nerve cell membrane contains ion pumps and channels that perform this function, leaving the inside of the cell negative and the outside positive. My last book challenged that view.[3]

An alternative view suggests that the charge separation arises from water. As we have seen, any water lying next to a charged or hydrophilic surface becomes EZ water; since cells pack charged surfaces so densely inside, most cell water is EZ water. With EZ water predominating, cell negativity could merely reflect EZ negativity.

Beyond explaining cellular negativity, the EZ hypothesis also explains the negativity of gels: gels commonly exhibit large negative potentials similar to those of cells; yet they lack any membrane that could pump ions. Thus, in this light, membranes seem almost irrelevant. If EZ water (rather than the putative membrane-based mechanism) powers the electrical activity of the cell, the observation that cells, including nerve cells, can often survive being sliced in half seems less paradoxical.[3]

If the EZ battery underlies the capacity to transmit signals, then eliminating the battery should eliminate the signals; the brain should never get the message. Local anesthetics do just that: the pain sensation never makes it to your brain. This action provides an experimental testing tool: if the EZ underlies signaling, then anesthetics should wipe out the EZ.

To test this hypothesis, we set up a standard EZ and added a local anesthetic. Clinical concentrations of either lidocaine or bupivacaine reversibly diminished EZ size in a concentration-dependent manner (**see figure below**). Local anesthetics do indeed wipe out the EZ, as we anticipated. This result might not surprise those who have read the old literature; after all, Linus Pauling, the legendary 20th century chemist, suggested something similar: an intimate link between anesthetic action and water.[4]

Beyond the mechanism of anesthesia, these observations imply something fundamental: an EZ basis for the cell's electrical features. It will be interesting to see whether additional research can confirm that the negative potential of the cell arises from the negative charge of the EZ.

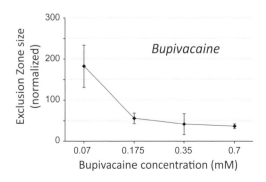

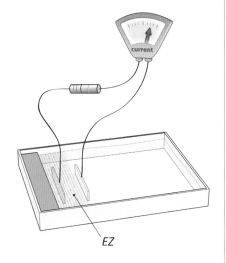

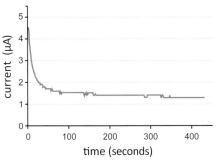

Fig. 5.7 *Current flow from separated charges in EZ and the water beyond. Current begins flowing immediately after immersing electrodes, maintaining a nonzero plateau value for an extended period of time.*

next to the pole-like EZs. Thus, we find an increase in protons next to EZs not only in standard situations (**Fig 5.3**) but also when the EZ projects deeply into the water.

In other words, charges separate wherever an EZ is present. Battery-like charge separation is an EZ fingerprint.

Harvesting Stored Energy from the EZ Battery

If the EZ's separated charges really behave like a battery, then electrical energy should be harvestable. Placing one electrode in the EZ, another in the zone beyond, and connecting those two electrodes through a resistor should produce current flow. That's what we found: the stored charges flow as current (**Figure 5.7**).

Hence, the separated charges are not merely incidental byproducts of EZ formation; they can be delivered to a load. The scenario resembles that of the common battery, but with simpler internal construction: here, an exclusion zone bearing negative charge sits next to a bulk water zone containing positive charge.

Think of it. Shortly after immersing a hydrophilic material into water, the EZ builds and charges separate. (The charge separation is no free lunch; we'll soon deal with the energy needed to drive the separation.) The separated charges have a strong tendency to recombine, but they remain separated because the EZ's dense lattice keeps those free charges from penetrating back into the oppositely charged EZ. The separation maintains the potential difference. The magnitude of that difference may reach only 100 to 200 mV, but the respective zones are nevertheless dense with charge; thus, the deliverable energy is substantial.

Water-based batteries of this kind exist wherever hydrophilic surfaces interface with water. That's virtually everywhere. In the cell, for example, the densely packed materials expose hydrophilic surfaces that order the surrounding water into EZs (see box); hence, cells contain numerous nanobatteries. Water batteries should also exist in aqueous suspensions and solutions, because EZs surround the suspended particles or dissolved molecules (see **Fig. 5.5**). Even water's containers can nucleate EZ-based charge separation. All of these scenarios create batteries, which are manifestations of water's fourth phase.

These batteries may seem unexpected to those trained within the framework of conventional thinking. However, we will soon see that this simple concept has immense explanatory power. Its potential will unfold as we deal with numerous water-related phenomena, ranging from osmosis all the way to ice formation.

Charge Carriers and Work Production

Understanding how the stored energy in such batteries might get delivered requires bearing in mind the species carrying the charge. That species is zone dependent. In the exclusion zone, electrons carry the charge; the electrons reside in the electronegative oxygen atoms distributed throughout the EZ lattice. The larger the number of oxygen atoms, the larger the number of electrons.

Those electrons can move easily from point to point throughout the lattice. Any such charge movement amounts to current flow, and we have confirmed the existence of current flow. **Figure 5.7** shows current flowing perpendicularly to the EZ planes. Charges can also move parallel to the EZ: the electrical conductivity measured parallel to surfaces that ordinarily nucleate EZs is 100,000 times higher than the conductivity measured through bulk water.[5] Hence, lattice electron charges can move easily in all directions, just as physicists know they move through n-type semiconductor lattices.

Beyond the EZ lie the positive charge carriers. Those carriers are nominally free protons; however, the actual carriers in water are hydronium ions — positively charged water molecules. This occurs because free protons seek out negativity, and negative sites abound in water's electronegative oxygen atoms. Negative sites are everywhere. Thus, protons quickly latch onto the nearest water molecule to form hydronium ions (H_3O^+). Hydronium ions carry the battery's positive charge, intermingled among ordinary water molecules.

Positively charged water is packed with potential. Since like-charged molecules repel, the hydronium ions disperse to remote

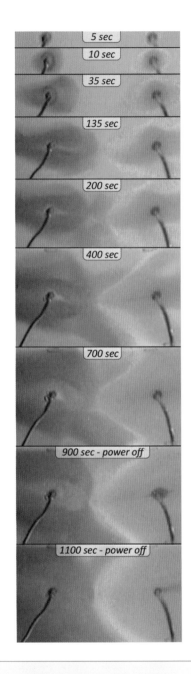

Fig. 5.8 *Time course of pH-dye distribution as current flows between wire electrodes immersed in a water bath containing pH-sensitive dye. Orange corresponds to low pH, purple to high pH.*

locations; this creates a liquid flow. Further, any remotely planted negatively charged sites should draw those hydronium ions, motivating additional liquid flow. Later, we will see how such attractive and repulsive forces constitute a primitive driver of natural water movement.

In other words, both the EZ's electrons and the bulk water's hydronium ions have considerable potential for doing work. Electrons can shift through the EZ lattice for delivery to contiguous sites hungry for electrons. Hydronium ions can drive flows and can also drive reactions requiring positive charge. Hence, both of those charged species can deliver abundant work-producing energy.

Efficient Energy Extraction

As you have just seen (**Fig. 5.7**), electrical energy can be extracted from water by placing electrodes into the water battery's oppositely charged zones. A related question asks whether we can extract energy efficiently.

This question arose out of a conversation with Andrey Klimov, a Russian colleague who first introduced me to the subject of water electrolysis. Andrey suspected that electrolysis might create long-term energy storage in water (similar to what we later found in the EZ system). We wondered whether electrolytically derived energy might be readily extractable.

In the simple electrolysis experiment, platinum electrodes are inserted at two locations in a water-filled chamber. A few volts (DC) are imposed between those electrodes. At first nothing obvious happens, but if you raise the voltage to a high enough level, you can see gas bubbles form on the electrodes. At lower voltages, no bubbles appear; nevertheless, electrical current still flows between the electrodes. Hence, charge must move into and out of the water.

To determine what might take place during this charge transfer, we added pH-sensitive dye. Something did happen (**Fig. 5.8**). Near the cathode, the color change indicated high pH; near the anode, it indicated low pH. The difference was easily six pH units — a proton concentration difference of a million times. The respective color zones spread progressively, and before long large regions of the chamber were one color or another. The chamber's water looked like a color chart: one region positively charged with low pH, another negatively charged with high pH (**Fig. 5.8**).

When we disconnected the power supply, the color blocks remained in evidence for some tens of minutes.[6] We would expect immediate charge annihilation if the positive and negative charges could combine, but the separation persisted for a considerable time. Meanwhile, you could extract current from the two electrodes. The charges remained concentrated in those two zones.

To determine how much of that charge could be extracted, we designed a more quantitative experiment (**Fig. 5.9**). We positioned plate electrodes at the respective ends of a rectangular chamber. A few volts were applied, and the dye color built into two semi-rectangular blocks, as shown. Then we disconnected the power supply. We found, once again, that we could draw current. We could draw it from the same electrode pairs used for charging or from a fresh electrode pair positioned at any of a series of points straddling the chamber's center-line. We could recover up to 70 percent of the input charge.[7]

The separated colors ordinarily persisted for tens of minutes, implying that the charges do not easily recombine. Probably those charges reside in EZ-like matrices — for the respective zones (**Fig. 5.8**) could be manipulated (by a permanent magnet) to shift or rotate within the chamber without substantially changing their shapes. The zones behaved like pieces of fabric. Embedding the charges in structural matrices evidently keeps the charges from recombining.

In fact, we confirmed the presence of structured matrices. Do you remember the 270 nm absorption peak that characterizes EZ matrices (**Fig. 3.12**)? Both zones showed a 270 nm absorption peak — the negative zone strongly and the positive zone more weakly. Thus, the stored charges reside in structural matrices, and that accounts for their long-term persistence.

We have shown that EZ batteries can capably supply charges. They can store charges for substantial periods of time; further,

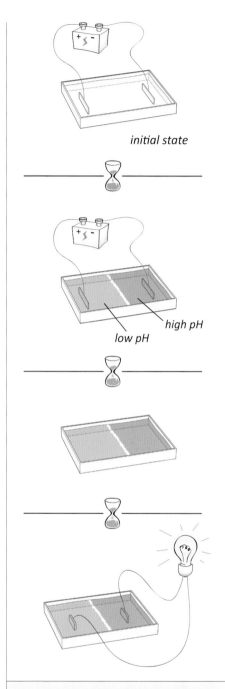

initial state

high pH

low pH

Fig. 5.9 *Time course of pH-dye distribution as current flows between wire electrodes immersed in a water bath containing pH-sensitive dye. Orange corresponds to low pH, purple to high pH.*

they can deliver a significant fraction of that charge. In subsequent chapters, we will see how that charge can provide energy for driving diverse processes, ranging from chemical reactions to hydraulic flows. Indeed, *the EZ battery could be a versatile supplier of much of nature's energy.*

Summary

Aqueous regions next to hydrophilic surfaces contain exclusion zones. Those EZs separate charges. The separated charges constitute a battery (**Fig. 5.10**).

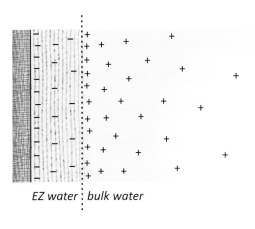

EZ water ⋮ bulk water

Fig. 5.10 *Diagrammatic representation of the EZ water battery. Hydrophilic surface at left. The separated charges are deliverable.*

One pole of the battery is the EZ, typically negative as a result of abundant oxygen atoms. The other pole lies in the bulk-water zone just beyond the EZ; it typically comprises positive hydronium ions, which can disperse freely according to the rules of electrostatics. Drawn towards negativity, many hydronium ions accumulate near the EZ boundary.

While the charge separation mechanism may now seem evident, the maintenance mechanism is not. Like your cell phone battery, the water battery will slowly run down as opposite charges trickle back together. The EZ battery, too, will need recharging. Since nature lacks wall sockets, some other source of energy must be at hand to do the job.

That source had eluded us for several years — until a chance discovery finally set us on the right track. We turn to that next.

6 Charging the Water Battery

It was Jim's casual, easy-going nature that led us to strike gold — or at least to find the equivalent: a source of energy virtually free for the taking.

My postdoctoral fellow Jim Zheng and I had been scratching our heads trying to figure out what energy kept the EZ charged. We seemed unable to find an answer. You had to first build and charge the EZ; then you had to maintain its negativity in the face of positive ions eager to penetrate and annihilate that charge. Thus, energy was needed not only for separating charge initially, but also for maintaining that separation in the face of inevitable attrition.

As for the buildup, we had a lurking suspicion that the culprit might be the "surface energy" present at material interfaces. However, surface energy didn't seem to fit: energy from the surface could reasonably build the layer closest to the surface, but EZs can build to hundreds of thousands, or even a million or more layers. How could some feature lodged at the surface act over so vast a distance? Something else seemed responsible.

As for maintenance, some energy had to sustain the charge separation once it was established. Maintenance seemed inexplicable without invoking some kind of continuous feed of energy in order to counter natural attrition. But the source of this energy feed remained unclear at the time — at least to us.

The first hint of the energy source came serendipitously from Jim succumbing to an ordinary human need. Laboratories are like other human workplaces: as evening approaches and hunger pangs grow pervasive, dinner bells sometimes chime almost audibly, and people will occasionally skip the standard cleanup. That happened one evening to Jim. He simply left the chamber on the microscope stage, turned off the microscope lamp, and ambled home for dinner.

Fig. 6.1 *Exclusion zone adjacent to Nafion.* Top: *control.* Bottom: *following several minutes of exposure to light.*

When he returned the next morning and turned on the microscope lamp to have a look, the exclusion zone had diminished to half of its former size. Within a minute or two, the EZ returned to its original size. It was as though the microscope lamp could reinvigorate the exclusion zone. Something about light seemed to matter (**Fig. 6.1**).

In retrospect, the role of light should have been obvious. When I raised this energy question during an undergraduate class presentation, a hand shot up and, half questioning, half making a declarative statement, one student blurted out, "Light?" He was spot on. The answer came so easily to that student (whose talents we quickly exploited in our laboratory); but for us, the answer had taken several years to figure out.

By the time of that class, we had managed to ascertain with some certainty that the responsible agent was light. I should be clear: by "light" I mean not only the visible part of the electromagnetic spectrum but also the ultraviolet and infrared portions. The vehicle of energy supply was radiant electromagnetic energy — which the water absorbs and uses for building the EZ and maintaining the attendant charge separation.

Light as Fuel

To account for light's mysterious expansionary effect, we first considered a potential artifact: a light-induced temperature increase. Incident light could heat the chamber and perhaps mediate the expansion. We quickly concluded that this was unlikely: the EZ's rapid growth had begun immediately after the light was switched on — well before the water in the chamber could have heated appreciably. Subsequent experiments confirmed that conclusion: even after the five-minute exposure that produced substantial expansion, the temperature rose by only a trivial amount.[1] Apparently, the effect of light was nonthermal: photons somehow donated their energy toward EZ growth.

That was an exciting moment. It appeared that sunlight could supply the energy needed for building order and separating charge; *the environment itself could do the job.* Imagine: energy from the sun could power the water battery in much the same way that the sun's energy powers photosynthesis. Wow!

Descending from those lofty heights of excitement, we asked the obvious question: which wavelengths of light bear responsibility for powering EZ growth? Ordinary microscope lamps (and sunlight) generate a broad range of wavelengths; the range encompasses visible light as well as ultraviolet and infrared light. We wondered whether certain wavelengths might work more effectively than others.

To answer this question, we shone lights with differing wavelengths into the experimental chamber. For the source of light, we used light-emitting diodes. LEDs emit light at specific wavelengths, ranging from the ultraviolet through the visible to the infrared. Using those LEDs one at a time, we directed the incident light onto the experimental chamber, which contained a strip of Nafion immersed in water. The water contained microspheres. We wanted to see how effectively the exposure to each of those wavelengths could expand the exclusion zone.

The results confirmed that wavelength did matter.[1] **Figure 6.2** shows what happened to the size of the exclusion zone after a five-minute exposure to light at each of a series of wavelengths. The incident light was weak enough that, by the end of the exposure, the chamber temperature never rose by more than 1 °C. The vertical axis shows the expansion — a ratio of 2, for example, indicates that the light exposure doubled EZ size.

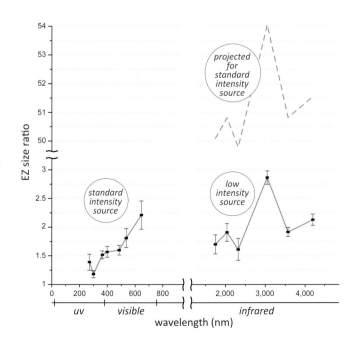

Fig. 6.2 *Effect of incident light wavelength on EZ growth. EZ size ratio (ordinate) refers to EZ size at the end of a five-minute exposure to light, relative to its size before adding light. For technical reasons, data on the right side were obtained with low-intensity light sources. Intensities similar to those used for obtaining the data on the left would have elevated the graph, as estimated by the dashed curve above.*

The figure shows that all wavelengths drove EZ expansion, but some wavelengths were more effective than others. Ultraviolet (including 270 nm) was least effective, visible light more effective, and infrared the most effective, *particularly at 3,000 nm,* which surprised us at first. Later, we realized that the 3,000 nm wavelength is the one most strongly absorbed by water. That means that the most strongly absorbed wavelength is the one that most effectively drives EZ growth — a rather satisfying correlation.

We also found that longer exposures and higher intensities could expand the EZ even more. The figure above was obtained using five-minute exposures. With longer exposures at the same intensity, for example, we could easily produce EZ expansion of five to ten times. Turning off that extra light brought the EZ back to its normal size within tens of minutes.

I should explain the mysterious dashed curve at the top of **Figure 6.2**. The infrared sources available for these experiments were feeble: they produced 600 times less intensity than the visible and UV sources. Thus, the right-hand series of data points surely fall below the levels they would have reached for IR light sources as intense as the visible light sources. How far below is uncertain. The dashed curve attempts to correct for that disparity; it offers one defensible estimate for the curve, assuming equivalent IR light sources.

Clearly, *infrared wavelengths dominate.* Ultraviolet plays almost no role. (We will deal later with the consequence of all the UV energy absorbed by the EZ.) Visible light plays a moderate expansionary role. And infrared (IR) wavelengths are by far the most effective for building EZs.

IR energy was probably the critical factor in Jim's accidental overnight experiment. When Jim casually extinguished the microscope lamp at the day's end, he reduced incident IR, which in turn diminished the size of the exclusion zone. When he turned the lamp back on the next morning, thereby raising the IR to the previous day's level, the EZ promptly returned to its former size.

Jim's accident also led us to investigate more systematic IR reductions (**Fig. 6.3**) — by inserting a chamber with fully developed EZ into an insulating container (a so-called dewar). In the

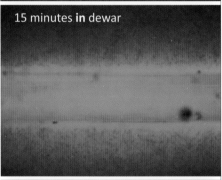

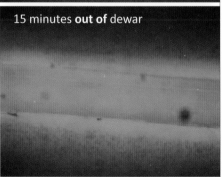

Fig. 6.3 *Reducing incident infrared energy diminishes EZ size. Extracting the chamber returns the EZ size to its original value.*

same way that a thermos can keep chilled drinks cold by blocking incident IR, dewars block IR even more effectively. After 15 minutes in a dewar, the chamber's EZ diminished to about half of its former size. Withdrawing the chamber returned the EZ to its normal size within some minutes. So the infrared effect works in both directions: increasing infrared expands the EZ, while decreasing infrared diminishes the EZ.

Think of what this implies (**Fig. 6.4**). Since infrared energy is most effective for building EZs, and IR is ever present, the fuel for building EZs is always available. The fuel comes free.

Fig. 6.4 *Even in the dark, infrared energy is freely available.*

In contrast to visible light, which can vanish at the flip of a switch, infrared light is difficult to shut out — IR cameras have no problem capturing images of rolling tanks or crowds of people even in complete darkness (**Fig. 6.5**).

Fig. 6.5 *Infrared image obtained in darkness. Brighter colors denote relatively high intensities.*

Even your room emits infrared. Your home's exterior walls absorb radiant energy from the sun and re-emit this energy at different wavelengths; in turn, your inside walls emit plenty of IR, whether the lights are turned on or off. Infrared is always present. Think of it as nature's gift — free for the taking.

Incident Energy May Dissociate Water Molecules

How could energy from light build the exclusion zone?

Light is versatile. Beyond producing images, light achieves many wondrous things because photon energy readily converts into other forms of energy. Examples:

- incident light of one wavelength converts to another wavelength, producing fluorescence;

- light powers the vibrational energy that drives Brownian motions (Chapter 9);

- light releases electrons in semiconductors to produce the photoelectric effect;

- light catalyzes reactions; and

- light separates charge in photosynthesis.

Given the versatility of light at converting energy in so many ways, finding that light drives EZ buildup should not come as a total surprise. This buildup of order and its attendant separation of charge merely constitutes another in a series of light-induced transformations — nothing particularly outlandish. In fact, light-induced ordering has already been experimentally demonstrated in other systems.[2] Hence, light-induced EZ buildup is neither exotic nor weird. The challenge lies in establishing the underlying mechanism.

A logical clue to that mechanism comes from the fact that EZ buildup can arise from energy sources other than just light. For example, we found that one such source, ultrasound, could drive

EZ growth. We applied 7.5-MHz ultrasound, similar to that used for imaging embryos. In response, the EZ typically narrowed, possibly as a result of the induced mechanical shear of molecules rubbing against one another. When we turned off the ultrasound, however, the exclusion zone immediately enjoyed a stunning regrowth: it could expand to five or six times its initial size before ultimately returning to pre-exposure levels. Evidently, the acoustic energy somehow affected the water, spurring a delayed EZ growth — just as incident light could produce EZ growth.

Finding that diverse agents can build exclusion zones makes it seem unlikely that light builds EZs by *directly* splitting the water molecule. Any such splitting would likely arise from a narrowly defined wavelength band, whose energy would resonantly shake apart the water molecule. Instead, we observe effectiveness over a wide range of wavelengths both within and beyond the optical spectrum. Further to this point, IR photons have low energy compared to, say, ultraviolet photons — so low that physicists consider IR completely unable to split the water molecule. Yet those infrared photons are the most effective EZ builders.

It appears that incident energy produces a subtler effect than direct water molecule breakup; more likely, incident light merely enables the breakup. The separation of charge would then take place at a later stage.

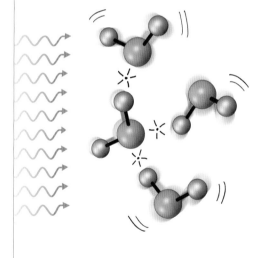

While the nature of that subtler effect remains uncertain, a reasonable speculation is that the energy separates water molecules from one another (**Fig. 6.6**). That is, the absorbed energy loosens intermolecular connections. Speculating much beyond that seems unproductive, for bulk water's structure itself remains a mystery. We know that constituent molecules must somehow stick to one another; otherwise water would be a gas rather than a liquid. As for how they stick, some scientists argue for transient intermolecular links, while others argue for ordered clusters that link with one another because of quantum mechanical effects (see **Chapter 2**); those ordered clusters might reorganize to build the EZ, as suggested in a recent proposal.[3]

Fig. 6.6 *Input energy may dissociate water molecules from one another.*

Thus, incident energy might weaken the links between water molecules or clusters, freeing water molecules for new "social" opportunities. That would constitute step one in the EZ buildup process.

Assembling the Exclusion Zone

Step two should involve some kind of assembly. Water molecules dissociated from one another must assemble onto the growing EZ lattice. Let us suppose that several layers of honeycombed sheets are already in place and ask how recently freed water molecules might assemble onto the outermost EZ layer to build the next layer.

The water molecule contains negative and positive charges slightly separated from one another. Those separated charges should each draw toward opposite charges lying on the lattice's exposed surface (**Fig. 6.7**). The molecule would then assemble onto the lattice. One by one, water molecules can settle in a similar fashion, thereby growing the next honeycomb-sheet layer.

While this process may seem straightforward, an inconvenient impediment arises: the hydrogen atom dangling from the lattice. **Figure 6.7** shows that once a water molecule joins the lattice, one of its hydrogen atoms dangles loosely. If that hydrogen remains, then the next sheet cannot build. Thus, lattice growth can continue regularly only if those dangling hydrogen atoms dissociate. Those offending protons must be cleaved.

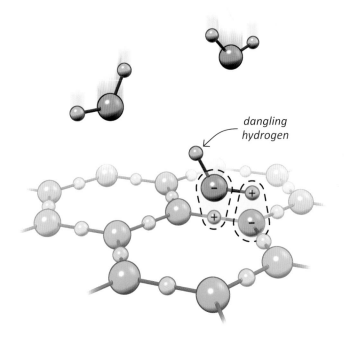

dangling
hydrogen

Fig. 6.7 *Building the exclusion zone. Freed water molecules draw toward the exposed EZ layer because EZ surface charges attract opposite charges on the water molecule. Following attachment, a hydrogen atom dangles.*

That cleavage can occur naturally. To understand how, consider the water molecule's electron clouds (**Fig. 6.8**). Oxygen's negative electron clouds point toward positivity (**a**). For the molecule lying in isolation, those clouds point toward the hydrogen nuclei; the resulting OH bonds keep the water molecule intact. This tidy scenario gives way as the water molecule settles onto the EZ lattice (**b**): as the molecule settles in, oxygen's electron cloud shifts at least partially toward the lattice; that cloud is the glue that helps bond the molecule to the lattice.

For the dangling hydrogen, however, that electron shift is jarring. The glue that formerly bonded that hydrogen to the oxygen to make a water molecule has vanished; the glue has shifted allegiance to the lattice. You might say the dangling hydrogen has come unglued — cast off as a lonely proton (**c**).

Fig. 6.8 *Electron cloud shifts as water bonds to lattice. The shift frees the dangling proton.*

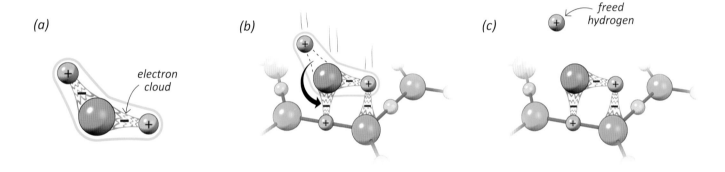

(a)   electron cloud

(b)

(c)   freed hydrogen

The process of casting off can also be viewed in terms of energy change. As each newly settled water molecule falls into place on the lattice, energy is released. This release occurs because oppositely charged entities lying at a distance from one another have plenty of potential energy, but as they merge, that potential energy gets released to the system. The scenario resembles separated magnetic poles, which surrender potential energy as they come together. Here that surrendered energy accomplishes something: it cleaves the dangling hydrogen atom.

So, looking through a structural lens or an energetic lens leads to the same result: proton cleavage. The lost proton carries positive charge, which is now separated from the negative EZ lattice. Effectively, the water molecule has split itself apart in order to grow the lattice.

By this mechanism, free proton charges build just beyond the EZ's growing edge (see **Fig. 5.3**). Some of those charges, repelling others, will rapidly diffuse into the bulk water. This diffusive action might have functional significance: if each and every one of those positively charged entities were to remain at the EZ boundary, the interface would clog; bulk water molecules would no longer gain access, and the EZ might quickly stop growing.

That's not the full story. As I mentioned earlier, free protons are actually short-lived species. Being free agents with positive charge, those protons will seek out anything electronegative nearby, in much the same way that many teenage boys will seek out any female nearby — almost any candidate will do. For the positively charged proton, the commonly available attractor is the water molecule's electronegative oxygen. The proton will latch onto that oxygen, creating an H_3O^+. That so-called hydronium ion is nothing more than a positively charged water molecule — an entity with significant potential, as I will show, for explaining all kinds of water movements.

So, when we talk about the dynamics of free protons, we are really discussing the dynamics of free hydronium ions. The hydronium ions are the long-lived species. They are the species that diffuse.

The main point, however, is that proton separation from the water molecule is a *secondary* event; it takes place as the water molecule settles onto the growing EZ lattice. The absorbed radiant energy that drives the entire process does not directly split the proton from the water. That radiant energy may merely loosen the bulk water structure, freeing individual water molecules to build. The act of latching onto the lattice then releases the water molecule's dangling proton into the bulk water, where it tends to form a hydronium ion. By these processes, the exclusion zone continues to build, and the water battery continues to charge.

An issue worth pondering is how a negatively charged lattice can keep adding more negativity. Adding negativity to negativity seems counterintuitive. However, that is not exactly what happens. The species actually added to the negative lattice is the neutral water molecule. The water molecule becomes attracted to the honeycombed EZ water because its negative and positive charges approach complementary positive and negative charges of the lattice. Those opposite charges are strongly attractive because of their close proximity, so the molecule sticks. Only later is the positive proton cast off in an energetically favorable way, leaving the lattice more negative. This stepwise process allows high concentrations of negative charge to build.

This process of building does not continue forever; eventually it terminates. Some pole-like projections may sprout from the EZ and continue to grow, but the main body of the EZ eventually attains a relatively stable value. A question arises: why does EZ growth cease? And, if incident light diminishes, how and why does the EZ shrink?

Exclusion Zone Disassembly

Here, as everywhere, we find the forces of nature at work. Ordered structures left alone will eventually become disordered. This gain of entropy is a fundamental feature of thermodynamics. It's kind of like your room: it can be messy in countless ways, while only few arrangements seem neat and tidy. To get it that way requires energy (**Figure 6.9**). Unless you put in that energy continually, your room will inevitably become as messy as, well … mine.

Fig. 6.9 *Building order requires considerable energy input. Creating the original mess requires much less energy.*

And so it is with the exclusion zone. Order cannot persist without a continual supply of energy. The separated charges will slowly recombine, and order will give way to disorder. The exclusion zone's outer reaches will wear thin like an eroding beach. That's what happened in that accidental overnight experiment: the EZ narrowed because a major energetic input was switched off; when that input was restored the next morning, the EZ grew back to its former size.

To understand what governs EZ size, then, we need to consider the balance between energy-dependent growth and the natural tendency to decay. When those two processes balance, the EZ attains a steady size. Growth was just explored, and growth-limiting factors such as surface roughness and degree of hydrophilicity were dealt with earlier. On the other hand, we've barely touched on the process of decay: how exactly does the EZ erode?

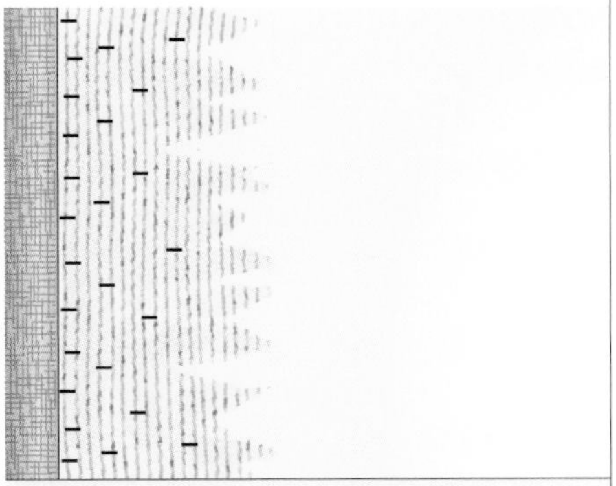

In order to answer that question, we need to consider the outer reaches of the exclusion zone, where you might expect attrition to take place. There, the electrical potential measures closer to zero, which likely means that some cleaved protons remain embedded in the lattice and/or the lattice is relatively more open (Chapter 4); see **Figure 6.10.**

A loose lattice implies easy molecular penetration. The most likely candidates for penetration are hydronium ions, for their positive charges are ineluctably drawn toward the high negativity of the EZ's inner reaches. So, hydronium ions invade the valleys between the EZ peaks.

This invasion has consequences. Having penetrated, these positive ions will be quickly captured by flanking negatively charged EZ molecules. The result is the merger that I alluded to earlier: an H_3O^+ combining with a lattice-structural unit (OH⁻), which yields two water molecules (**Fig. 6.11**). This erosive action loosens the EZ's hexameric structure.

Fig. 6.10 *Jagged outer reaches of the exclusion zone. Hydronium ions penetrate the valleys between the mountain peaks because of attraction to negative charges.*

Fig. 6.11 *Natural erosion of exclusion zone. Combining a hydronium ion with one EZ structural unit extracts that unit from the lattice, resulting in two water molecules.*

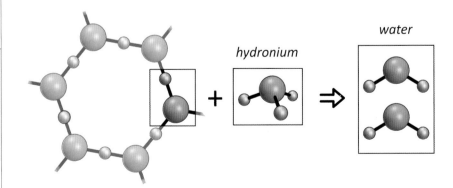

hydronium

water

So we are back where we started: an element of the EZ lattice has returned to water, and the system has taken a step back. The system reaches a steady-state size when creation and destruction become balanced, i.e., when energy-driven EZ production balances natural EZ attrition.

That balance can shift as ambient conditions change. In acidic water conditions, the ample hydronium ions in bulk water should continually chip away at EZ mass, tilting the balance towards a smaller EZ. We have confirmed this experimentally: sufficiently acidic pH does diminish EZ size. Salts erode the EZ similarly. Consider NaCl: While the Cl^- component can combine with H_3O^+ in the bulk to yield $HCl + H_2O$, the positive Na^+ can invade the negative lattice, and go on to create NaOH by extracting a lattice OH^- unit. The EZ erodes and adds a water molecule to the bulk water. Wherever the lattice is open, positive ions of any sort can enter and cause EZ erosion.

In sum, the exclusion zone retracts by a process that largely reverses the way it builds. It builds from attracted water molecules joining the lattice and casting off protons, many of which immediately become hydronium ions. The lattice retracts when hydronium ions invade lattice openings and extract EZ units to yield water. The balance point will depend on how much energy enters the system: more intense incident energy yields larger exclusion zones, while less intense incident energy yields smaller ones.

Free Radicals

No process is perfect, including the EZ's buildup-attrition dynamic. Central to that dynamic is the OH^- structural unit. The EZ builds by locking those OH^- units into the lattice one at a time, whereas it retracts by releasing those units one at a time to the bulk water. The process is thus reversible — more or less. It is fully reversible provided that hydronium ions happen to be at hand to sop up each freed OH^- and create water; then the system will go back to where it started.

Suppose, however, that hydronium ions are in short supply locally and therefore unavailable to do the job. That could happen, for example, if a negatively charged site, located at an appreciable distance from the EZ, has drawn the hydronium ions away. Then, no partner is available for neutralizing the lattice OH⁻ unit, so the cycle cannot complete. Likewise, cycle disruption could arise from perturbation of the EZ itself: Suppose some electron-hungry process draws off some of the EZ's negative charge, leaving the released lattice unit devoid of its usual negativity. Again, the cycle can't proceed. Issues of this nature could upset the default situation.

In such cases, the reversible cycling scheme outlined above will not be so neat and tidy. Instead of yielding water, the withering process will yield various alternative forms of oxygen, which will then populate the bulk water. The natures of those alternative oxygen species will depend on the nature of the deviation.

These alternative oxygen species are generally known as free radicals, or sometimes, because of their high reactivity, reactive oxygen species (ROS). The most common one, the superoxide radical, comprises two oxygen atoms with a single negative charge. Another, the OH radical, bears no charge. Still another is H_2O_2, or hydrogen peroxide. All of them contain oxygen, and theoretically all of them can arise from exclusion-zone breakdown.

The high reactivity of these species can cause problems. High reactivity implies instant binding to many substances, and that binding can potentially alter those substances. In living systems, these reactions can induce toxicity: for example, the superoxide radical can be a potent killer of microorganisms.

Unsurprisingly, nature makes every effort to scavenge these radicals in order to avert any such consequences. Thus, every cell in your body contains a scavenging enzyme called superoxide dismutase, or SOD. SOD neutralizes emerging superoxide radicals almost as rapidly as they form. This enzyme's omnipresence had remained something of an enigma. On the other hand, if free radicals occur as natural byproducts of EZ dynamics, then the enzyme's ubiquity becomes understandable: since exclusion zones form practically everywhere, SOD should be practically everywhere as well.

Life in the Depths

Recognizing the character of these energetic processes can help unravel some of nature's mysteries, and I cannot resist mentioning just one of them here: why the bottom of the deep sea hosts so many living creatures. Those depths lack not only dissolved oxygen, but also light. Organisms can neither breathe nor photosynthesize. That should make life impossible, but paradoxically, it flourishes (**Fig. 6.12**). Each time a deep-sea sample is taken, scientists identify more and more species. Even bacteria that are obligate photosynthesizers have no difficulty thriving in such lightless environments.[4]

Fig. 6.12 *Life at the deep bottom. Brine pool eel, a creature of the deep sea floor, taken in the Gulf of California. Courtesy NOAA. Wikipedia commons.*

The energetic processes outlined previously may explain this enigma. Although extreme depths certainly lack visible light, they suffer no shortage of infrared light. IR energy radiates from the earth itself, and particularly from thermal vents that line the ocean's floor. IR wavelengths build EZs and separate charge. The charge-separation process bears much similarity to the initial steps of photosynthesis, where water molecules split; hence, bacteria and other deep-sea creatures may exploit this mechanism to gain energy.

Furthermore, oxygen is not at all absent near the ocean floor. The IR-built EZs contain plenty of oxygen; therefore, oxygen can come from the EZ. My colleague Vladimir Voeikov refers to this process as the "burning" of water. Dissolved oxygen may be absent, but as long as exclusion zones exist, ample oxygen remains available for fueling life's processes.

So, despite the bleakness of the deep-sea environment, we can understand life's abundance at those extreme depths — energy and oxygen are in plentiful supply. In any case, this deep-sea discussion was opened only to whet your appetite for the delights of energetics: the underlying energetic processes are fundamental not only for deep-sea life but also for much of nature. We will explore the broader implications of those energetic processes in the next chapter.

Summary

The exclusion zone builds from light energy, particularly infrared. Infrared energy is available even with the lights turned off. Acoustic energy can also do the job. These energies plausibly dissociate bulk-water molecules from one another, freeing them to build the EZ. Drawn toward the growing EZ by charge attraction, the freed water molecules assemble onto the lattice. This results in EZ growth and attendant separation of charge. In this way, the interfacial battery gets charged.

The EZ assembly process resolves the previous chapter's quandary about how EZ charges can pack so densely. Negative charges repel; hence, the EZ should rightly fly apart. However, electron clouds glue each new element onto the growing lattice, ensuring integrity of that lattice. Those electron clouds can be likened to the tabs that hold puzzle pieces together (**Fig. 6.13**). Net repulsions notwithstanding, the pieces remain firmly interlocked.

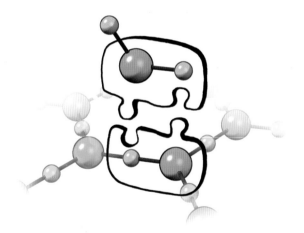

Fig. 6.13 *Adding elements to the lattice. Repulsion notwithstanding, elements remain bound together by the "puzzle-piece" interlocks.*

The EZ may begin to disassemble if it lacks an adequate supply of energy. Separated charges inevitably trickle back into the lattice. When that happens, elements of the EZ matrix degrade into the same water molecules that bore responsibility for EZ buildup. Without enough incident energy to counter this erosion, the growth process reverses, and the water battery discharges.

When conditions for reversal fall short of ideal, oxygen radicals may form instead of water. Those radicals can be nasty. To avert their destructive power, biological systems take special measures: they provide abundant enzymes to sop up the radicals as fast as they form. Self-preservation seems one of nature's prominent attributes.

Perhaps you've wondered what happens to all the energy that the EZ builds. Does the EZ finally go up in smoke? Or does something more useful take place? The next chapter addresses this question. It asks whether a lowly glass of water can actually use the stored energy to do work.

7 Water: The Engine of Nature

My colleague Vladimir Voeikov has a passion for experimentation. During my recent visit to his weekend dacha outside Moscow, Vladimir proudly pointed to the windowsill, where an array of light-exposed beakers full of water was under study. He then pointed to the garden below, where another experiment was in progress, this one executed by his wife and daughters, decked out in their best garden duds.

Gardening is relatively new to the Voeikovs. Only recently had they acquired their dacha. Russians seem to have a genetic passion for growing vegetables, and the Voeikovs were eager to try their hand. Their immediate neighbors had been gardening for generations, yet Vladimir's plants stood fully one third taller. This mildly embarrassing achievement arose not out of any special gift or unusual dedication, for the Voeikovs' thumbs were not noticeably greener than most. Something else was responsible for their success.

Fig. 7.1 *Vladimir Voeikov, in his office, ruminating on his next experiment.*

Vladimir claimed it was the water. His professional life includes time spent within a 200-km radius of Moscow University searching for and testing natural waters that are "energized." This term may sound vaguely new-agey; but such energized water has become a central component of a medical regimen now legendary in the Moscow area. Not surprisingly, Vladimir uses that same water for growing his plants.

Can water really contain energy? Early pioneers, including Viktor Schauberger and Rudolph Steiner, provided ample evidence that water could store and deliver energy; and contemporary scientists are beginning to revisit this possibility. On the other hand, conventional wisdom says otherwise: a closed bottle of water sitting on your tabletop is considered in equilibrium with the environment; the environment might heat up and transfer energy to the water, but other than this slow thermal process, no obvious mechanism should allow the water to receive and store energy, let alone deliver it. Water is water — dull as a doorknob and hardly a vehicle for storing any kind of energy beyond low-grade heat energy.

Or is it?

Water as an Energy Converter

When water absorbs light, the absorbed energy builds structural order and drives charge separation. That stored potential energy is harvestable: charge separation can produce electrical current (Chapter 5); and structural order can drive cellular work.[1] These conversions confirm water's ability to store and deliver potential energy.

On the other hand, can we be sure of the generality of this transduction? Even if the previous chapters had made a flawless case for energy storage and release in water, that case relies on a single line of experiments from a single laboratory. That's insufficient to establish generality. Hence, we look elsewhere for evidence that water absorbs energy from outside and converts it to useful work. Let me begin with the work of a legendary Italian scientist.

Piccardi's Marathon

On an otherwise uneventful flight from Seattle to Frankfurt, I immersed myself in something memorable: a book written by the distinguished chemist Giorgio Piccardi (**Fig. 7.2**). A colleague had recommended this classic, but for the life of me, I could not imagine how a book entitled *The Chemical Basis of Medical Climatology*[2] could possibly bear on the subject of water and energy. Once I began reading, however, even a sputtering airplane engine could not have drawn me away.

Piccardi was intrigued by the statistical variability of his experimental results. One day a reaction might take two seconds to complete, the next day 2.5 seconds, the subsequent day 1.8 seconds, and so on. To learn more about the source of this variability, Piccardi and his colleagues started a series of daily experiments extending over some twelve years (except for a brief hiatus during World War II). They conducted nearly a quarter of a million experiments. The central question: why did reaction rates vary from one experimental run to the next? Every experimentalist knows that this happens, but few understand why.

To obtain an answer, Piccardi explored a variety of diverse reactions in parallel. All of them involved water. The reactions included a simple

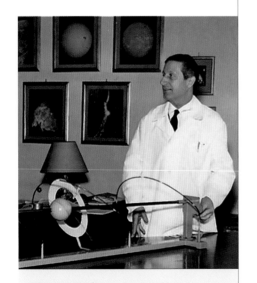

Fig. 7.2 *Italian scientist Giorgio Piccardi (1895 – 1972).*

chemical precipitation, the formation of a polymer, and a phase change involving the freezing of supercooled water. The end points of these reactions were clear enough that their timing could be measured with precision. Each day, with practically religious adherence, Piccardi and associates would carefully combine reagents and record the reaction times. Other parameters, such as temperature and pressure, remained constant.

All experiments were carried out in pairs. One of each pair was placed inside a metallic Faraday-cage shield or beneath a horizontal shield; the other was prepared identically but left unshielded. The shielding against electromagnetic waves was a critical feature of the experiments. The team would then measure reaction times in the respective situations, for the variety of reactants.

Reaction times varied from one day to the next, as expected. Piccardi noticed, however, that the mean values of the reaction times depended on whether the samples had been shielded or unshielded. The difference was consistent. This led Piccardi to conclude that, beyond local and known variables, some feature of the environment must influence the reaction rates. And, because those differences showed up consistently for the several different reaction types, the scientists concluded that the environmental influence had to be general.

Piccardi further concluded that water had to be involved. Since water was the only common element among the diverse reactants, it appeared to Piccardi that the water must have absorbed some kind of environmental energy, which then influenced the reaction times.

Although the nature of that environmental energy never became clear, the researchers found tantalizing clues for its origin. They observed repeating cycles. From December through January each year, the degree of variation in reaction times sharply dipped, but then began to increase around March, reaching a maximum during June and July. The same cycle repeated each year. They also noted other recurring phenomena. For example, reaction times varied with the natural periodicities of solar activity, and especially with sunspots and solar flares — clearly implying that the sun's energy played a role.

Following systematic analysis with extensive controls, Piccardi concluded that the only plausible explanation was that the radiant energy absorbed by the water must have played a role in these reactions. As incident energy varied, so did the reaction times. The cycle periods were pivotal: they implied that the energy could come from the sun and possibly also from the cosmic background of space.

Piccardi's work generated a significant following. That following included a special "Piccardi Group" established within the framework of an international scientific society. Although members of the group ultimately dispersed, one prominent Russian investigator has pressed on for more than four decades with experiments extending Piccardi's work.

More Enigmatic Oscillations

Simon Shnoll and his colleagues deviated somewhat from Piccardi's approach. Like Piccardi, they studied the timing of biochemical reactions; however, they also studied the timing of seemingly unrelated phenomena, including radioactive decay counts, and certain gravitational behaviors.

From such timing data, they constructed timing histograms — graphs showing the likelihood of possible timings. Shnoll focused on the histograms' "fine structure," i.e., the squiggles on the curves. Generally, those squiggles showed little similarity from one curve to the next. However, data pairs obtained at intervals of 24 hours, 27 days, and 365 days showed remarkable similarity. The similarity was not merely eyeballed; analysis was made using objective methods that left little possibility of chance. From the periodicities observed, Shnoll concluded that all phenomena studied must be influenced by geophysical or cosmo-physical sources — much the same as Piccardi concluded.

Both Shnoll's and Piccardi's results emphasize the role of energies beyond those commonly considered. If such "exotic" energies have an effect then they must first be absorbed; and, since both investigators' experiments involve water, water would seem the likely absorption target. Shnoll's results go a step further, provocatively implying that these energies might also be absorbed in nonaqueous physical systems.

Still More Oscillations

Further evidence for radiant energy absorption in water comes from Vladimir Voeikov, whom I mentioned earlier. Voeikov studies the light emitted from aqueous solutions. The intensity of that emitted light oscillates with the daily cycle (**Fig. 7.3*a***). The experiment illustrated in the figure took place in a light-tight chamber with a controlled temperature; hence, the oscillation exceeds any plausible variation arising from external temperature fluctuation. The emitted light oscillation was evidently the effect of some kind of radiant input beyond visible light, which had been blocked. The radiant energy evidently varied with the diurnal cycle. This implies an influence of solar energy.

Another such recording revealed more. In **Figure 7.3*b***, note the sharp upward inflection near the beginning of the curve. Suspecting some experimental fluke, Voeikov checked and found that the upturn coincided exactly with the onset of a local lunar eclipse; this suggested that cosmic energy could impact the light output. That cosmic impact was evidently strong enough to overshadow the daily oscillation.

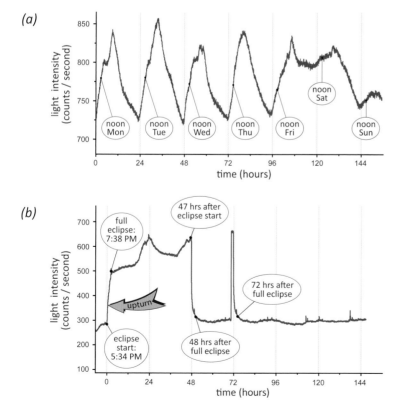

Fig. 7.3 (a) *Light emission recorded from water containing bicarbonate ions and enhanced by dissolved luminol. Note the periodic intensity variation.* (b) *Similar to (a) but recorded during an eclipse of the moon.*

The correlation between the lunar eclipse and the sharp deflection might have been fortuitous; however, certain features of the curve implied otherwise. At 24 hours following the eclipse, a modest downward recovery began; at 48 hours, a sharper recovery occurred; and at 72 hours, the record shows another sharp downturn. While these transients are not easily explainable, their propensity to occur at 24-hour multiples following the eclipse seems unlikely to be coincidental. Hence, the conclusion drawn here is the same as that drawn by Piccardi and Shnoll: incident radiation from some cosmic source seems to affect the water.

In sum, water absorbs energy from the environment. No other plausible interpretation comes from the periodicities seen in all of these studies. Thus, the previous chapter's evidence showing the influence of incident light does not stand alone; other experimental approaches confirm that incident radiant energy affects water. Incident radiant energy evidently impacts many features of water, ranging from increasing the speed of reactions all the way to generating light.

These results make clear that a sealed flask of water resting on a table is not a closed system; it is open to the environment. The water behaves much like a plant sitting next to the flask. The plant is an open system: it uses the radiant energy that falls incident upon its surface. The same appears true of the flask of water. This similarity should not surprise us since plant cells, after all, comprise mostly water.

Energy Transformation: The Engine of Nature

If water absorbs radiant energy, then what happens to all of that energy? Can the water continue to absorb energy in endless amounts? Or, must the water process that energy in some way?

A helpful analogy is the inflation of a balloon. Increasing internal pressure confers potential energy to the balloon. If you let go, the balloon flutters about, releasing that potential energy kinetically. The energy you invested converts to another type of energy. On the other hand, if you keep inflating without release, then the balloon will eventually rupture: all of the energy will release in one calamitous pop.

Water continuously absorbs radiant energy from the environment; however, the water doesn't explode like the balloon. Therefore, some sort of continuous energy release must be built into the system. In the balloon analogy, you might say that the balloon must be letting off a continuous stream of air (**Fig. 7.4**) — a pressure release akin to a long bout of flatulence.

How, and in what forms, does the energy get released from the water?

You have already seen a few examples, above. I will show you that those examples belong to a much larger set. Energy can be released from water in many forms: optical, physicochemical, electrical, and mechanical. In other words, water acts as a machine that transduces input radiant energy into many kinds of output energy.

(i) Optical Energy Output

I already mentioned Voeikov's demonstrations of light emission. He has recently extended those experiments to show emission occurring over long periods of time. Voeikov filled containers with water, adding modest amounts of bicarbonate, peroxide, and a small amount of luminol for light amplification. He then sealed the containers and used a photomultiplier to examine light output over time.

The results were unexpected. Following the initial recording of light output, Voeikov stowed the containers in dark cupboards, testing them only occasionally. After well over a year of dark storage, the same sealed water flasks continued to deliver light. The intensity did decrease slightly, but the flasks continued to emit light for unfathomably long times. The light refused to go out (**Fig. 7.5**).

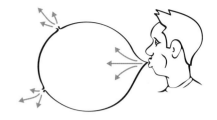

Fig. 7.4 *Relief valve. A continuous release of energy assures that the system never overloads and explodes.*

Fig. 7.5 *Aqueous solutions can produce practically eternal light output.*

Fig. 7.6 *Salt water exposed to electromagnetic energy. The solution catches fire.[3]*

You might expect light output from some chemical reactions — but continuing for over a year? Either some magic is at play or the aqueous solutions must continuously absorb incident energy and convert that energy into the practically unending photon-energy output that is observed. No need to belabor the point: the water solution acts as a light bulb, delivering photonic energy practically endlessly, with no obvious source other than the energy stored within the water.

A more in-your-face demonstration of optical energy output from water comes from water on fire. A demonstration is shown in **Figure 7.6**. The image shows a tube of salt water held by a bracket. The salt water is exposed to microwave or radio-frequency energy, and voila! — light and heat.[3] You can see it on video:[w1] a vivid demonstration that water can convert input energy into light.

(ii) Physicochemical Work

Turning from light, consider a second output: physicochemical energy. Envision a beaker filled with water, containing suspended particles such as microspheres. The suspension is nominally uniform at first, but after several hours, something seemingly mysterious happens: the microspheres gravitate toward the beaker's periphery, leaving a vertically oriented cylinder near the center that is microsphere free (see **Fig. 1.3**). The microspheres are said to "phase-separate," leaving one region rich with microspheres and the other devoid of them.

Systems left alone commonly tend toward disorder, not order. Entropy, after all, is time's arrow. But the system above seems to move from disorder to order: initially scattered randomly throughout the suspension, the microspheres eventually crowd near the beaker's periphery. Such crowding is akin to a group of people initially mingling in casual conversation and subsequently asked to crowd into half the space. That will not happen spontaneously; it requires determination and energy.

The same is true of the observed microsphere crowding: some kind of energy must drive the crowding. Radiant energy is the obvious agent, and **Chapter 9** will confirm this. Here, however, the driver is of less consequence than the product — the almost deterministic reorganization toward more condensed arrangements. To reorganize, the microspheres must move through a viscous medium, and that requires work. Separations involve work.

Separations occur in various other kinds of suspensions, and work is done in each case. Such work might fit more naturally into the later section on mechanical work, but phase separations are ordinarily classified as physicochemical phenomena; hence, we include them here. Either way, the observed particle movements provide evidence of a class of energy output over and above optical energy output.

(iii) Electrical Work

We can also extract electrical energy from water (**Fig. 7.7**). As you have already seen in **Chapter 5**, placing electrodes into oppositely charged zones of the water battery produces electrical current. Whether such energy production can compete with existing technologies remains uncertain; nevertheless, the water battery does demonstrably produce electrical energy from input radiant energy.

In fact, we have also obtained electrical energy from zones of opposite charge that were electrically induced (**Fig. 5.9**). By inserting electrodes into the oppositely charged regions, we were able to extract

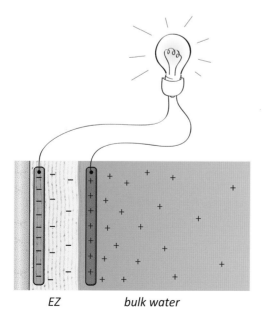

EZ   bulk water

Fig. 7.7 *Electrical energy generated from electrodes placed in the EZ and in the zone beyond.*

substantial energy — practically as much as the electrical energy used to build those charged zones.

Thus, water can deliver electrical energy. Imagine using water to run your cell phone! That prospect should not come as a complete surprise, for water-based batteries can already produce enough electrical energy to power a clock (**Fig. 7.8**).

(iv) Mechanical Work

In the context of mechanical work, I refer to water movement, or flow. Producing flow requires energy input: if you transport water uphill, you expend energy; you might even develop a sweat. Even if you drive the water through a horizontal tube, you still need to expend energy in order to overcome molecular friction or viscosity. Driving flow of any kind requires energy input.

Now, suppose neither you nor some other supplier of energy is on hand to drive the flow. Then the energy must come from the water itself. That's what I would like to illustrate next: flow production in the absence of an obvious driver — other than the subtle energy stored in the water. I present three examples.

(a) Tubes. The most dramatic example is the flow through hydrophilic tubes (**Figure 7.9**). To see this, drop a 1-mm length of Nafion tubing into a small chamber of water, taking care to ensure that the water fully permeates the inside of the tube. Then make sure the tube lies flat at the bottom of the chamber. To track any flow, add some microspheres or a blob of dye.

Fig. 7.8 *Water clock. The underlying principles differ from those that electrochemists might surmise (see Chapter 12).*

Fig. 7.9 *Practically incessant flow occurs through hydrophilic tubes immersed in water.*

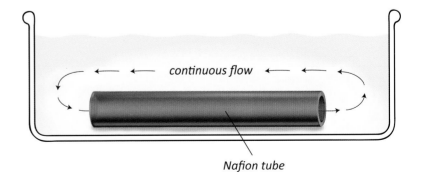

continuous flow

Nafion tube

You might expect that nothing much would happen, but something does happen: after a few minutes of chaotic startup, you will see steady flow running through the tube, much like the blood running through a vessel. Its direction is unpredictable from one trial to the next, but once it gets going, it persists with little diminution for as long as an hour;[4] and when steps are taken to ameliorate the effect of protons accumulating in the chamber, the flow can persist for more than a day.[5] If you reorient the tube during this period, the flow maintains its direction relative to the tube.

We have observed this kind of flow not only through Nafion tubes but also through cylindrical tunnels bored within various gels. The results are similar. Thus, rather than being specific to any one material, this flow phenomenon apparently occurs due to the materials' hydrophilic nature. Tubes made of hydrophobic materials generate no flow. Evidently, some kind of local interaction between hydrophilic surfaces and water drives the flow.

While the detailed driving mechanism is not fully worked out, some aspects are clear (**Fig. 7.10**). An EZ demonstrably builds just inside the tube *(a)*; we can see it. The EZ generates a buildup of hydronium ions in the core of the tube *(b)*; we can measure it. When the hydronium ion concentration builds sufficiently, those positively charged water molecules must begin escaping at one end of the tube or the other, to the fluid outside. That escape initiates the flow *(c)*. The escape draws fresh water into the other end of the tube. The incoming water gets protonated, which perpetuates the flow.

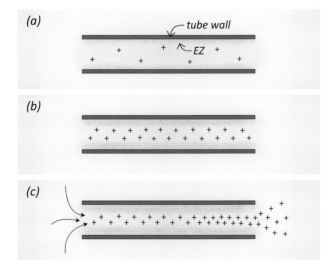

Fig. 7.10 *Mechanism of intratubular flow. The key element is the hydronium-ion buildup in the tube core, and its escape to the water outside.*

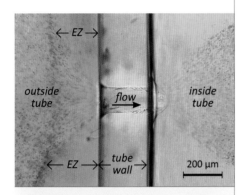

Fig. 7.11 *Inward flow produced by hole punched in the wall of a Nafion tube. View is from top of chamber. The inwardly flowing microsphere suspension moves towards the two open ends of the tube.*

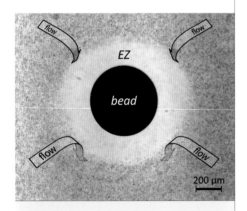

Fig. 7.12 *A gel bead (top view) positioned in a chamber containing water and microspheres generates a consistent flow pattern around the bead.*

Once flow begins, light enhances that flow.[5] White light enhances flow in an intensity-dependent manner; ultraviolet light can enhance the flow by as much as four to five times. Thus, the energy driving this intratubular flow evidently comes from light. The light presumably releases protons, which drive the flow.

(b) Holes. Intratubular flow is not the sole manifestation of "spontaneous" water flow. In another example, consider a small hole punched in the wall of a submersed Nafion tube. Water instantly flows into the tube through the hole (**Fig 7.11**). This inward flow, observable by adding tracer microspheres, has a surprisingly high velocity. Although the velocity eventually diminishes, flow is unexpectedly persistent over long periods of time. Some type of energy must drive this flow, and various observations again point to free protons,[6] here attracted to the negative charge inside the tube.

(c) Beads. A third example of seemingly spontaneous flow occurs around gel beads. Place a half-millimeter gel bead on the floor of a small experimental chamber. Then cover the bead with water so that the top of the bead is just barely submerged. Add some microspheres to help track the flow. The nature of the flow is surprising: in the fluid layer nearest the water's surface, the water consistently flows toward the bead from all directions (**Fig 7.12**). As the water approaches the bead — or the bead's exclusion zone — it turns downward, toward the chamber floor; it then continues on an outward course away from the bead. The flow effectively circulates outside the bead — again propelled by the positive hydronium ions' attraction to the negative EZ.

We have seen this circulatory flow with different types of beads and different types of chambers. Invariably, the flow continues without fatigue for many hours, at least until the microspheres have settled to the bottom of the chamber so flow can no longer be tracked. Some kind of energy must drive this persistent flow (presumably with a top-to-bottom gradient), and the obvious candidate is the radiant energy absorbed by the water.

When viewed in the context of standard energy paradigms, these three flow regimes may seem mysterious; they resemble perpetual motion machines. In theory, the flows could be rationalized if they arose from side effects, such as thermal gradients, but each study carefully probed for and ruled out such artifacts. The flows seem slightly less mysterious when viewed in the context of the EZ, for the EZs lodged next to all of these material surfaces release protons. Even minor proton (or hydronium ion) gradients will drive flows, because charge gradients always want to even out. *Charge gradients are powerful drivers of all kinds of flow.*

Which brings us to a tangential point: what happens to the flows in the absence of a material that directs them? Suppose no tube, hole, or sphere is present to organize and direct these flows. What happens then? Radiant energy continues to enter the system; but how does all that energy dissipate in the absence of something to organize the flows?

Perhaps you guessed that movements might still occur, albeit undirected. Random displacements do occur ceaselessly in water. Known as Brownian motions, these motions are consequential for physics and chemistry; we will deal with them in Chapter 9, where we explore whether Brownian motion might be driven by incident radiant energy.

For now, it should suffice to appreciate that absorbed electromagnetic energy has mechanical consequences. It may drive flows of all kind, which constitute work output.

Photosynthesis-like Energy Conversion

Water evidently performs all kinds of work, ranging from chemical and optical to electrical and mechanical. The potential energy driving that work comes from charge separation, which in turn comes from the action of absorbed radiant energy. That stored energy serves as an intermediary, which can drive all kinds of work or energy output.

This chain of events bears a striking resemblance to photosynthesis. In photosynthesis, absorbed radiant energy from the sun drives energy outputs not unlike those we've just seen: chemical energy (metabolism); mechanical energy (bending); flow (in plant vessels); and, in some organisms, even light. Input radiant energy produces varied types of work. You see examples all around in green plants and microorganisms.

Do You Photosynthesize?

Plants do it; bacteria do it; various single-celled organisms do it. Photosynthesis is a process so effective for species lower on the phylogenetic tree that one cannot help but raise the question whether nature retained that process as it advanced to more complex organisms. I'm not necessarily suggesting that you photosynthesize, for the common end point of photosynthesis is the building of organic substances; for us, those substances come easily from food. However, your body might take the first step in the photosynthetic process: using incident light for separating water's charge. The separated charges would then become useful for driving various physiological processes.

A plausible process is the driving of blood through capillaries. We've just seen that light drives flow through hydrophilic tubes; in similar fashion, light could propel flow through your superficial capillaries, which are merely hydrophilic tubes. Plenty of light penetrates into your body to do the job — as you can confirm by shining a flashlight through the palm of your hand; if you do this in the dark, you can easily see the light emerging from the other side. Thus, penetration of light into your body could plausibly help drive capillary flow.

Blood flow can use all the help it can get. In healthy young adults, red blood cells can be larger than the capillaries through which they pass: 6 to 7 μm vs. 3 to 5 μm, respectively. Those red cells demonstrably contort as they pass.[w2] Think of pushing a partially deflated soccer ball through a toilet drainpipe and you've got the picture: doing so requires substantial pressure, even with low

friction. By contrast, routine measurements show almost no pressure drop across the capillary bed, most of the drop occurring over the relatively larger arterioles. Thus, capillaries do not behave as the high-resistance conduits we'd expect. Some energy could assist the heart in driving the flow, and that energy could easily come from the radiant energy that your body absorbs.

Thus, Mother Nature might not have disappointed us. In her vaunted wisdom, she might have retained a mechanism from plants and bacteria, adapted to convert light into flow and other types of work in animals. The first step may be generic: light transduction by a hydrophilic substance next to water. While you might never sprout new shoots or bend toward light, you might nevertheless exploit the same transduction mechanism used so effectively by the plant on your windowsill.

The first step in the photosynthetic process is the splitting of water. The splitting into positive and negative components is mediated by light-absorbing chromophores lying next to the water. That scenario resembles the one considered here: a hydrophilic surface lying next to water. In both cases, light induces water-molecule splitting. It seems tempting, therefore, to think of the light-induced water splitting of photosynthesis as similar to the light-induced water splitting associated with the EZ. This means that photosynthetic chromophores are merely specific embodiments of the more generic hydrophilic surfaces. Light incident on chromophores and water splits the water, just as light incident on hydrophilic materials and water splits the water — the former probably more effectively.

If chromophores are nothing more than specific embodiments of generic hydrophilic materials, then photosynthetic reaction centers ought to show the 270-nm absorption peak characteristic of an EZ's presence. Blue-violet and red absorption peaks are textbook material; however, professionals studying photosynthesis also find an impressively large 280-nm absorption peak. That 280-nm peak is commonly dismissed as a nuisance attributed to protein contamination, but one wonders: could that (nominally) 280-nm peak be the same as the 270-nm EZ signature peak? If so, that would strongly indicate the presence of EZ water inside the photosynthetic reaction centers. The EZ would reside next to the chromophore, and the gestalt would be quite similar to what happens next to generic hydrophilic surfaces.

In sum, the first step of photosynthesis could be a specific embodiment of the more generic process outlined in these pages: light-induced splitting of water and separation of charge. *EZ-based charge separation may be a generic first step of the photosynthetic pathway.*

The Balance of Energy

Multiple examples have shown that the EZ stores potential energy for subsequent use. However, not all of that stored energy remains available. This is because some energy gets radiated back to the environment, a fraction of it in the form of heat. Infrared cameras, and sometimes even simple thermometers, can detect that heat. So the output consists not merely of work, but work plus radiant energy.

A convenient way of summarizing this concept is in the form of an equation:

radiant energy in = released energy or work + radiant energy out   (1)

Equation *1* refers to a steady state. It does not apply to transient states: for example, if input radiant energy abruptly increases and begins expanding the EZ, then an extra term would be needed to account for that transiently stored energy. For the steady state, however, equation 1 should suffice.

The equation expresses the steady-state energy balance. It emphasizes that transduction is less than 100 percent efficient, since some of the absorbed energy returns to the environment from which it came. Only the unreturned fraction gets converted to useful energy or work. On the other hand, this unreturned fraction is marvelously capable: it provides energy for any number of processes.

Summary and Prospects

We conclude this section of the book by reflecting on where the past chapters have brought us, and where the newly revealed features of water might lead us.

We first identified an unexpected feature of water. Next to hydrophilic surfaces, we found that water molecules organize into liquid crystalline arrays, which can project unexpectedly far from their nucleating surfaces. Like crystals of ice, these liquid crystals exclude many substances, ranging in size from macroscopic colloidal particles to submicroscopic solutes. The prominence of this exclusionary feature gave rise to the moniker "exclusion zone."

Exclusion zones commonly bear a negative charge, while the bulk water zones beyond them contain complementary positive charge. The two zones have different characters: the negatively charged EZ seems to comprise a semi-crystalline fabric built of stacked honeycomb sheets; the positively charged zone is featureless, containing hydronium ions free to disperse or flow according to the whims of electrostatics.

The energy required for building the EZ and separating charge comes principally from radiant sources. Infrared light seems particularly effective. It is also omnipresent and free for the taking. Acoustic energy seems also to show this capability, although details remain to be elucidated. All of these energies may act by dissociating bulk water molecules, thereby freeing those molecules so they can add to existing EZ layers like bricks on top of a partially built wall. The water molecules stick naturally, and in the process of sticking, they lose positive charges to the bulk water. In this way, charges separate and the water battery gets charged.

The energy flow implied by this process is unconventional in that the absorbed radiant energy does not merely degrade as heat, as generally thought (**Fig. 7.13**). Some of that energy can get converted into potential energy; that potential energy can be delivered in various energetic forms — including chemical, optical, electrical, mechanical, and perhaps other types of work or energy. In other words, there are two energetic pathways.

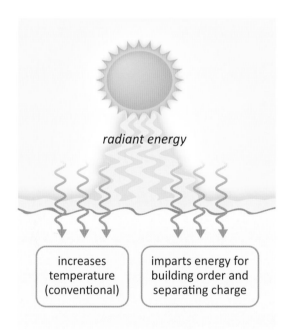

radiant energy

increases temperature (conventional)

imparts energy for building order and separating charge

Fig. 7.13 *Energetic pathways within water. The conventional pathway is heating; the newly revealed pathway is the creation of deliverable potential energy. This latter pathway may be the generic first step in photosynthesis.*

Water, therefore, acts as a transducer, absorbing one kind of energy and converting it into other kinds. The conversion may occur instantaneously, as in fluorescence, or it may get held in reserve for future use, such as perhaps for the growing of vegetables taller than those of your neighbor.

So we arrive at the second equation,

$$E = H_2O \qquad (2)$$

Equation 2 emphasizes that water and energy go hand-in-hand. Purists may decry the mismatch of units, for which I have no defense. Nevertheless, I think you understand what I mean: energy and water are closely linked. Wherever water exists, so does stored energy, and that stored energy can do all kinds of work.

Think of what this means. You put energy into water and you get other kinds of energy out (**Fig. 7.14**). Water is an energy converter — a liquid machine, if you will.

Envision how the workings of this machine could impact practically every feature of water. Consider, for example, water's heat capacity. Heat capacity refers to the amount of heat required to raise the temperature by a fixed amount. For water, heat capacity is larger than conventional chemistry would predict: the water heats up less quickly than the pot in which it sits.

Fig. 7.14 *Water transduces energy.*

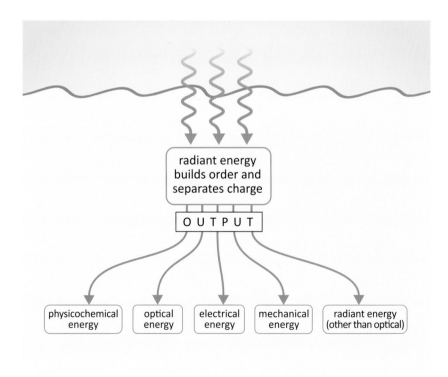

The reason for water's high heat capacity has been a matter of debate — but please refer back to Figure **7.13**. Radiant energy certainly raises water's temperature, but some of that radiant energy is drawn off to build structure. That is, only some of the input energy goes toward heating. As a result, water needs to absorb a larger than expected amount of radiant energy in order to raise its temperature by some amount.

Heat capacity illustrates only one of the many energy-related issues involving water; other issues range from the evaporation of warm water to the freezing of cold water. You might think that we understand those phenomena, but that's not the case at all. Many anomalies remain — which is another way of saying that we haven't a clue what's *really* going on. Numerous water-related mysteries remain.

For resolving these mysteries, an obvious starting point involves developing an understanding of what happens to water as you add or subtract energy — and that's where we're headed in the next section of this book.

We now begin using what we have learned about water to bring a broad range of natural phenomena into better focus. Our model of the EZ sheds new light on long-established concepts and — as I hope you will soon find — radically shifts how we understand our world.

SECTION III

What Moves Water Moves the World

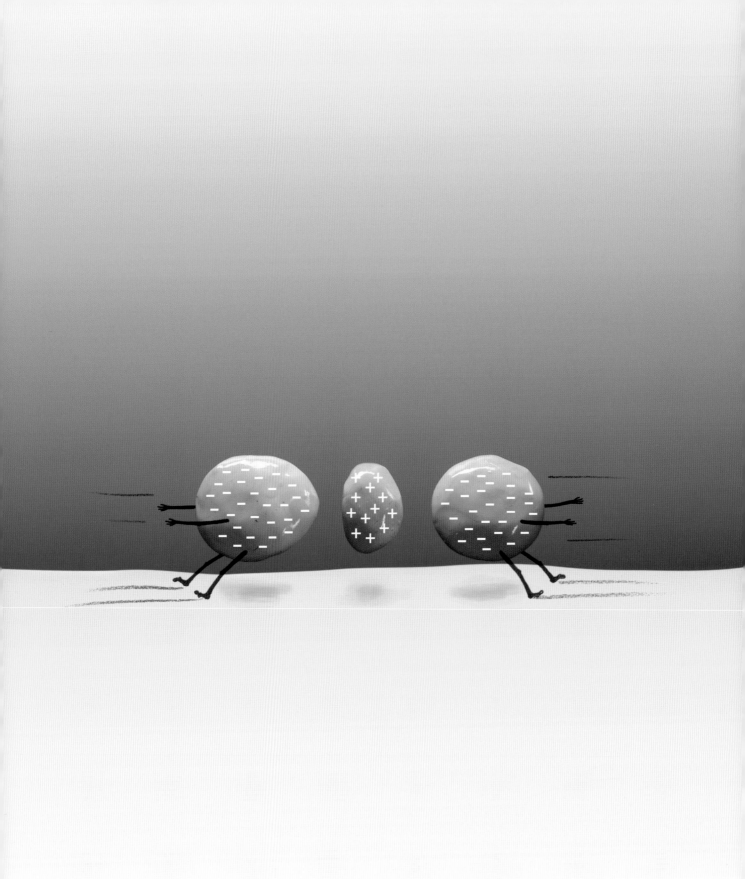

8 A Universal Attractor

If we can call anything in science fundamental, surely that must include the notion of charge: opposite charges attract, and like charges repel. Simple enough. So let me pose a question: Suppose that you extract a charged particle from your left pocket and a similarly charged particle from your right pocket. You place those two particles in a beaker of water, positioning them close enough that they "feel" one another's charge. What happens to the distance between those two particles?

When I ask this question during talks, typically no hand will rise. People sense a trick and fear that the wrong answer might publicly confirm their deeply rooted insecurities. A brave soul might eventually raise a hand and meekly suggest: "well…uh…the like-charged particles will obviously repel and move apart."

In fact, they move *toward* one another.

Before you rashly conclude that this author must be on some kind of drug, let me assure you that this paradox is not a hallucination. The phenomenon has been known for a century. Irving Langmuir, a figure significant enough to merit an eponymous physical chemistry journal, knew this phenomenon well.[1] The famous physicist Richard Feynman subsequently offered a sensible explanation that did not defy any fundamental tenets of physics.[2]

Richard Feynman (1918-1988)

So why would two like-charged particles want to approach each other? And what might that phenomenon imply for the rest of natural science?

Mechanism of Paradoxical Attraction

Feynman put forth a simple explanation for this attraction. In his inimitable style, he opined: "like likes like" because of intermediate "unlikes." In other words, like-charged entities attract (or "like") each other, because charges of opposite polarity lie in between. An intermediate positive charge will thus draw two negatively charged spheres closer together (**Fig. 8.1**).

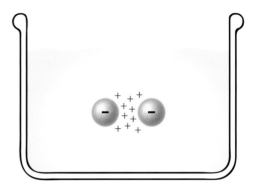

Fig. 8.1 *Negatively charged particles may attract if enough positive charges lie in between.*

We will see ample evidence in this chapter that "like likes like" really does work — extremely well. We will address the obvious questions: From where do those unlikes come? Why should they gather between the likes? Finally, what determines the end point of the attraction? This chapter will go beyond answering those questions to show that this mechanism provides a simple framework for explaining many paradoxical phenomena.

Had Feynman been in the audience when I asked my question, he would assuredly have blurted out the answer, but his answer would have been conjectural; at the time, we knew little about the source of those opposite charges. It took Norio Ise of Kyoto University to gather supporting evidence for Feynman's profound notion.

Ise studied colloids. Colloids are relatively homogeneous mixtures of particles and solvent — think of yoghurt, blood, or milk (**Fig. 8.2**). The particles are smaller than pebbles but larger than molecules; typically, they are on the micron scale. You can detect their presence because the particles scatter light and therefore lend opacity to the suspension. Seeing the individual particles requires a microscope; nevertheless, they are present throughout the colloid, thickening both the milk you drink and the blood passing through your veins.

Ise explored colloids made of microspheres and water — as simple as you can get. He found that, if he waited long enough after mixing,

Fig. 8.2 *Milk is a common example of a colloidal suspension.*

the particles would redistribute themselves into regular arrays called "colloid crystals." **Figure 8.3** shows an example. The figure illustrates the crystal's two most notable features: (i) the particles become regularly spaced; and (ii) the particles remain separated from one another. The separations may appear small, but the inter-particle distances are large enough to include lineups of thousands of solvent molecules.

Ise and his colleagues found that these crystals formed through electrical attraction.[4] Immediately after mixing, the microspheres dispersed throughout the solvent medium, but over time, they drew progressively toward one another, leaving behind microsphere-free voids surrounding the clusters (**Fig. 8.4**).

The story did not end there. Over time, the microspheres within those condensed zones underwent further condensation, finally turning into regular arrays like the one illustrated in **Figure 8.3**.

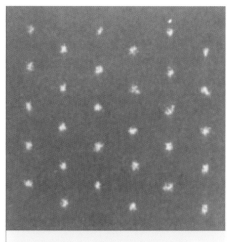

Fig. 8.3 *Eventual distribution of latex particles: 0.4 μm in diameter, 2% concentration in aqueous suspension.[3]*

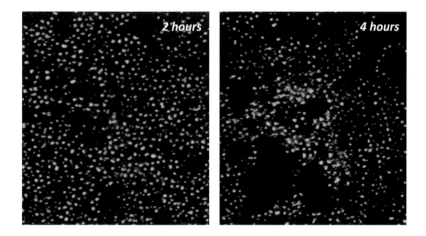

Fig. 8.4 *Microspheres draw together over time, leaving large voids that contain no microspheres. Frame width: 10 μm.[4]*

Ise and his colleagues went on to explore the effects of many variables on the drawing force.[5,6,7] In every case, the Feynman prediction fit: the results could be explained only if unlike charges lay between the particles of like charge. Ise and colleagues thus concluded that Feynman was on track. The paradoxical attraction did actually arise from something quite orthodox: attraction between *unlike* charges. No need to invoke anything even hinting at a violation of basic physics.

Fig. 8.5 *Norio Ise receiving Japan's highest science prize. Emperor and Empress at left. Courtesy Japan Academy of Science.*

For the elegance of his work, Ise received the highest science prize one can receive in Japan: dinner with the Emperor. Both the food and the conversation, I'm told, were excellent (**Fig. 8.5**).

Confirming Long-Range Attractions

While straightforward, the Feynman-Ise mechanism has not sat well with some physical chemists, who argue that any such attractions should be impossible. Their views are based on the so-called DLVO theory, which precludes long-range interactions. DLVO is the acronym drawn from the surnames of those responsible for that formulation: Derjaguin and Landau (Russian); and Verwey and Overbeek (Dutch). Thus, "DLVO."

The DLVO theory describes the force developed between two charged surfaces facing each another in a liquid medium. DLVO theory builds from the presumption that each charged surface will attract counter-ions from the liquid; these adsorbed ions will mask the surface charge, like a shroud masks a corpse (**Fig. 8.6a**, *left*). Thus, an observer sitting in the liquid at some distance from the surface would hardly "feel" the presence of that charged surface.

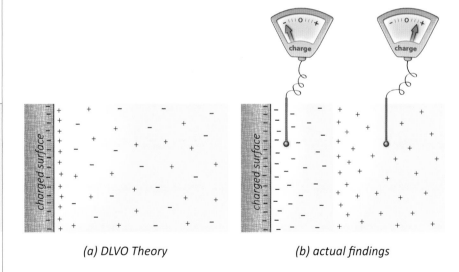

Fig. 8.6 *Contrasting charge distributions. (a) Charge distribution presumed for DLVO theory, where counter-ions gather next to charged surface. (b) Charge distribution measured experimentally.*

(a) DLVO Theory   *(b) actual findings*

That's why DLVO theory can't account for the particle attractions. Charged surfaces lying even a short distance apart cannot feel one another, according to that theory, because the counter-ions mask the charges; therefore, the surfaces cannot attract. The critical separation distance for nonattraction depends on various factors but is rarely more than a few tens of nanometers and more often is only a few nanometers. By contrast, Ise's observations show attractions even at separations 100 times larger, and in a moment, I'll show you attractions occurring practically on a *millimeter* scale. DLVO theory does not fit those observations.

DLVO theory has a fundamental problem. At its very heart, that theory presumes a distribution of charge that conflicts with the measured distribution. **Figure 8.6b** schematizes the experimentally measured charge distribution (see also Chapter 4). The figure shows a vast zone of charge extending from the surface to a substantial distance into the water; that is the EZ charge. DLVO theory predicts nothing of the kind (**Fig. 8.6a**): the counter-ion-masked surface charge leaves little or no net charge beyond the mask. Hence, the theory conflicts with the measured reality.

Physical chemists nevertheless adhere tightly to DLVO theory. Many therefore doubt that the observed attractions could possibly exist; and a few have carried the flag of doubt rather aggressively.[8] Ise has responded vigorously to each and every one of those challenges;[9,10] as far as I know, his defense remains unchallenged.

Nevertheless, it was tempting to see if we could confirm, and perhaps even extend, the Feynman-Ise attraction; to do so, we launched experimental tests. For particles, we used gel beads. These half-millimeter beads are galumphing masses by molecular standards — millions of times the volume of the microspheres used commonly by Ise and others. Their large size helped us see what was really going on.

We placed two like-charged beads on the floor of a small chamber containing pure water, positioning the beads at some distance from one another. Then we waited to see what might happen. Occasionally, the beads would move spontaneously toward one another, hinting at the anticipated attraction; more commonly, they would sit resolutely where they started. It occurred to us that the beads' tendency to stick to the chamber floor might obscure any evidence of attraction, so we introduced a protocol of lightly tapping the bottom of the chamber to free the beads. That strategy worked (**Fig 8.7**).

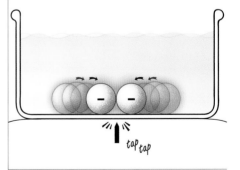

Fig. 8.7 *Experimental technique used to unstick beads from floor.*

With each tap, the beads visibly freed themselves from the floor and could therefore move about freely before settling down and sticking once again. This multiple-tap protocol let us track the distance between beads.

The results were clear:[11] the beads consistently attracted one another (**Fig 8.8**). Even beads initially separated by close to half a millimeter drew progressively closer. This attraction happened consistently, with both negatively and positively charged bead pairs. In fact, the positive beads stuck less to the surface and often attracted without any tapping at all.

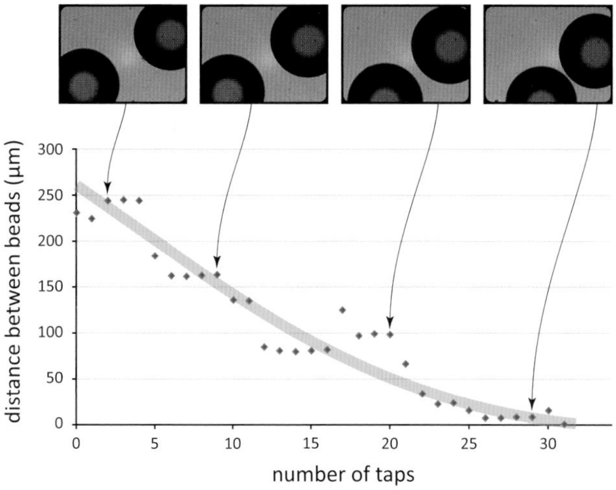

Fig. 8.8 *Bead separation as a function of time. Beads move progressively closer. Pictures above correspond to data points beneath.*

So the attraction was confirmed in the simplest possible system: two large beads in water. It seems that like likes like even from afar — a result that we found particularly pleasing.

The remaining challenge was to track down the unlike charges theoretically required for mediating the attraction. Could we be certain of their presence? Also, where might the unlike charges come from?

Confirming the Role of Unlike Charges

One possible source of unlike charges is the exclusion zone. EZs build next to hydrophilic surfaces by virtue of captured radiant energy. The surfaces can be flat or spherical (Chapter 4). The buildup process yields unlike charges beyond (**Fig. 8.9a**). Thus, a negatively charged sphere suspended in water should be surrounded by numerous positively charged ions.

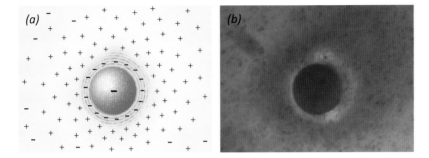

Fig. 8.9 *Spherical particles immersed in water. (a) Diagram showing surrounding charges. (b) Negatively charged bead (dark) immersed in water with pH dye. Lighter region around the dark sphere is the EZ, which excludes pH dye. Intensely red region beyond indicates low pH, or high concentration of protons.*

We have confirmed the presence of those positive charges experimentally. **Figure 8.9b** shows a gel sphere immersed in water containing pH-sensitive dye. Beyond the dye-free EZ, the intense red color indicates extremely low pH — a high concentration of hydronium ions. Thus, the two panels correspond nicely.

Now suppose that, instead of a single negatively charged particle, you have a pair; suppose further that those two particles reside not too far from one another (**Fig. 8.10**). Abundant positive charges surround each EZ-clad particle. The largest number of positive charges should lie between, because that intermediate zone contains contributions from both particles.

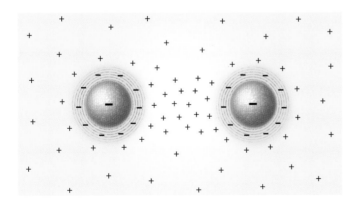

Fig. 8.10 *Expected distribution of charge when two spheres lie near one another. The positives in the middle do not significantly disperse because of their attraction to EZ negativity.*

The result? The spheres will move *toward* one another, i.e., in the direction of highest positive charge.

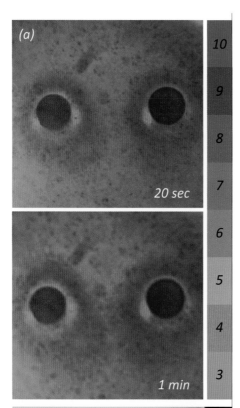

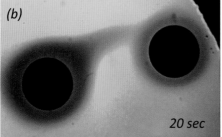

Fig. 8.11 *Unlike charges situated between like-charged spheres. (a) Dye shows region of low pH (numerous protons or hydronium ions) in between two negatively charged beads. At right, pH scale. (b) Region of high pH (negative charges) lying between two positively charged beads. Negative charges sometimes coalesced into bridge-like structures, as seen here. In all images, fluid drift moves the unlike intermediates slightly off center.*

Please note that this mechanism violates no laws of conventional physics. Like charges *do not attract*. Opposite charges do the attracting, just as you learned in freshman physics. Those opposite charges present themselves in highest numbers between the particles. Thus, plentiful charges are available for mediating the attraction, even at unexpectedly large particle separations (**Fig. 8.8**).

Can we actually detect those in-between charges?

With beads of macroscopic dimension, it became feasible to test for the charges' presence.[11] We tested this in two ways. First we used pH-sensitive dyes (**Fig. 8.11**): the dye color confirmed that, with two negatively charged beads, the largest number of protons occurred in between (*a*); the dye also confirmed that with two positively charged beads, the OH⁻ concentration (high pH) was highest in between (*b*).

The second test used fine microelectrodes. Positioned at points throughout the zone between the beads, those electrodes confirmed what the dye experiments showed: an electrical potential with opposite polarity to that of the EZ.[11]

Hence, the opposites required to mediate the attraction are indeed present. Feynman's hypothesis is validated.

Force Balance: Establishing an End Point

With a mechanism in place that explains the attractions, a question arises: when does the attractive movement stop? Observations show that the spheres generally stop moving before they bump into one another (**Fig. 8.3**); but why should that be? Why shouldn't the particles smash into one another?

The secret lies in a repulsive force: like charges repel. The repulsive force between the like-charged particles is weak when the particles lie far apart; there the attraction dominates. Then, as the negatively charged spheres draw closer to one another, the negative-negative repulsion increases. When that repulsive force increases enough to balance the attractive force, the movement should cease.

Such balance can yield either of two end points, and in our experiments, we have seen both. For typical micron-sized particles in water, the end point occurs when the particles remain some distance apart (**Fig. 8.12**, *top*). There, repulsive and attractive forces balance, yielding the standard colloid crystal seen in **Figure 8.3**.

For larger particles, such as the half-millimeter beads, the attraction may continue to outweigh the repulsion even as the particles touch. The reason lies in scale: the regions flanking the touch point remain separated by sizable gaps (**Fig. 8.12**, *bottom*), in which many opposite charges can dwell, sustaining the attraction. Meanwhile, the repulsion remains relatively weak because of the large separations between the negatively charged surfaces. With strong attraction and weak repulsion, attraction predominates even as the spheres literally bump into one another — at which point the attraction continues to hold them together. With multiple spheres, the bumping turns into clumping. Crystalline order persists, but with no separation between elements.

In any event, crystals form when repulsive and attractive forces equal one another. Abundant charges keep the system in tight balance, even in cases with large particle separation. Since opposite charges govern the balance, you might expect that incident light will affect the separation. Experiments have confirmed this:[12] increasing the light draws the particles closer together.

Solutions versus Suspensions

In dealing with colloidal particles, I have repeatedly used the term "suspension." What exactly is a suspension? And how do suspended entities differ from dissolved entities?

Physical chemists make a clear distinction between the two phenomena. Chemists consider molecules dissolved because they interact with the surrounding water; i.e., molecules possess "shells of hydration." Particles differ. According to the prevailing view, particles merely mix with the water, eventually settling to the bottom.

However, particles *do* interact with water: they demonstrably interact to form exclusion zones, which should qualify as "hydration."

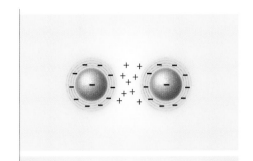

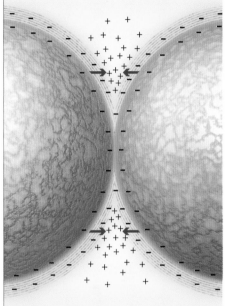

Fig. 8.12 *Stable points for smaller* (top) *and larger particles* (bottom). *End points occur when attractions balance repulsions.*

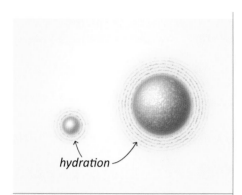

Fig. 8.13 *Similar hydration of molecule and particle implies that dissolution and suspension are similar in principle.*

In this sense, suspended particles are much the same as dissolved solutes (**Fig. 8.13**). The same governing mechanisms may apply to both. To appreciate the logic of this argument, think of an entity whose size lies at the boundary between particle and molecule. Is that entity suspended, or is it dissolved? Which set of rules should govern?

If these two phenomena demonstrate a single principle, on the other hand, then one shoe might fit all mechanistic feet. Prevailing interpretations might change. We'd hope that the more forbidding explanations that now fill chemistry textbooks might give way to simpler ones.

Implications

The main message of this chapter is that everything seems to attract everything else. Like attracts unlike, and we've just seen that like attracts like. In the latter case, the descriptive catch phrase "like-likes-like" may be cutely memorable, but I stress that the rules of conventional physics remain unbroken; physics still works as you learned. The point is that attraction is practically universal; at least in water, everything attracts everything else.

However, the mechanisms governing the two classes of attraction differ, especially in terms of energy. We take the attraction of unlike-charged objects as axiomatic: so far, no evidence has challenged the notion that positive attracts negative. Mediating that attraction requires no energy. In fact, when particles of opposite charge approach one another, they actually *release* potential energy associated with the separated charge.

The like-likes-like attraction, by contrast, is subtler. Here, the attraction ultimately arises from absorbed energy: absorbed radiant energy builds EZs and separates the charges responsible for mediating the attraction. The more intense the absorbed light, the more intense the attraction.[12] *So the like-likes-like attraction requires energy:* as long as the sun continues to deliver energy, like-charged entities will continue to attract.

Like-likes-like attraction is not restricted to colloidal particles suspended in water. EZs also occur in diverse polar solvents, including ethanol and acetic acid.[13] With separated charges, the like-likes-like principle should apply there as well. In theory, the principle could

Salt and sugar crystals represent extremes of the like-likes-like mechanism. To produce sugar crystals, you dissolve sugar (sucrose) in water and heat the solution in the presence of an immersed seed crystal. As the water slowly cools and evaporates, a crystal forms. This "hard" crystal is known as rock candy. Its many opposing charges keep it solid.

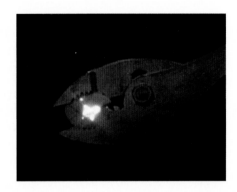

The presence of those charges can be confirmed by cracking the crystal in the dark (see figure). As the crystal breaks, the separated charges jump across the fracture, producing a discharge similar to lightning.

The sparks are popularly seen with individually wrapped wintergreen-flavored Lifesaver candy. Grab pliers, place the wrapped candy in the pliers, dim the lights, and allow your eyes to adjust in the dark. Then, crush the lifesaver crystal across the wide axis with the pliers and you will see the flash. (Opening the wrapper at that stage lets you enjoy the candy dust.) For a more personal encounter with science, open the candy and have your friend crunch the crystal orally: if your friend's mouth is dry enough, you'll see blue sparks.

apply not only in liquids, but also in any situation in which electromagnetic energy drives charge separation. That could happen from atomic to cosmic scales.

Some examples:

• *Atomic scale.* Consider gaseous hydrogen. Two proximate hydrogen atoms share electron clouds to form the gas. Thus, negative charges lie between two positively charged nuclei.

• *Physical scale.* An array of like-charged metallic balls placed on a gently shaken insulating table soon takes on two-dimensional order; and sure enough, opposite charges on the insulating substrate lie in between those charged metallic balls.[14] The situation resembles a two-dimensional colloid crystal.

• *Biological scale*. Consider freshly synthesized biomolecules, which self assemble into larger-scale structures, including filaments and vesicles. The assembly mechanism is incompletely understood. Might the like-likes-like mechanism draw those molecules together?

• *Organismal scale*. The schooling behavior of fish usually gets explained in evolutionary terms. However, a slimy, gel-like substance coats the fish surface. Gel-like substances create exclusion zones, with protons lying beyond. Could fish capitalize on the like-likes-like mechanism to help organize more efficiently?

• *Cosmic scale*. Charged plasmas dominate cosmic phenomena.[15] "Dusty" plasmas appear in the rings of Saturn, in the tails of comets, and in the numerous interstellar clouds that fill space. Particles of these dusty plasmas arrange themselves in ordered crystal-like structures. So similar are those plasma crystals to colloid crystals that they are often called "colloidal plasmas."

So, while the like-likes-like mechanism was originally advanced to explain the paradoxical behavior of colloidal particles in water, the principle's applicability could range from the atomic scale to the cosmic scale. That span includes phenomena we see every day but don't really understand — and never think to explain by the like-likes-like mechanism.

A couple of examples:

• *Clouds*. Consider the puffy white cloud in the clear blue sky (**Fig. 1.2**). Clouds are built of water droplets. Droplets of like charge should ordinarily disperse, but they obviously coalesce to form the individual clouds that we see. Coalescence could easily arise from the like-likes-like mechanism — opposite charges holding those droplets together to form discrete clouds.

If those droplets are regularly arranged as in a colloid crystal, then certain mysteries resolve. Imagine a beam of late afternoon sunlight hitting one of those clouds. Constituent droplets scatter the incident light in all directions. If those cloud droplets have achieved uniform spacing, then the light scattered at specific angles, depending on wavelength, will lie in phase. The result: a splendid rainbow.

- *Sandcastles.* Standing to protect us against the specter of invading flotillas, sturdy sandcastles (Chapter 1) may achieve their strength by the like-likes-like mechanism. Those castles are not built from sand alone; they also contain water, which enables the building of EZs around each sand particle (**Fig. 8.14**). Thus, protonated water molecules lie between the EZ-enveloped sand particles. Those unlikes constitute the glue that holds together the castle.

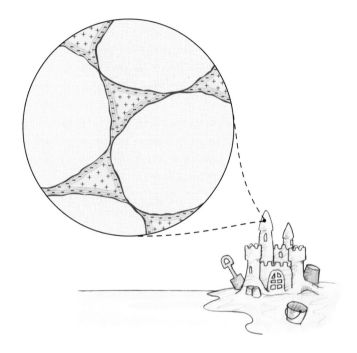

Fig. 8.14 *Like-unlike forces glue sand particles together.*

Let me speculate that the like-likes-like mechanism may be universal. At minimum, its existence nullifies the reflexive view that like-charges *must* repel—replacing that view with a more nuanced model of particle-particle interaction that includes attractions. If this like-likes-like principle proves extensively valid, then multiple foundations could crumble.

So don't be surprised to see the like-likes-like principle popping up in the chapters to come; the principle explains otherwise inexplicable phenomena. Potentially, *the like-likes-like mechanism may be foundational for all of nature.*

Summary

Charge envelops particles suspended in aqueous media. That charge builds as water transduces ambient energy into structured exclusion zones. The exclusion zones bear one polarity, while the water beyond contains charges of opposite polarity. Those opposite charges concentrate most between neighboring particles, explaining why the particles attract and move toward one another. The attraction occurs naturally.

Known as the like-likes-like mechanism, this attraction arises from the presence of unlike charges lying between the like-charged particles. Unlike attracts like. Hence, the rules of physics are not violated. At the most basic level, like still prefers unlike.

Think of the philosophical implications of this attraction. In Japanese culture (dating from the 11th century *Tale of Genji*), the way to bring two warring parties together is to put an attractive woman in between; that brings the men together (**Fig. 8.15**). Like-likes-like operates similarly: the intervening opposite creates the attraction. Thus, attraction is more or less universal: opposites attract and likes attract. If you allow some poetic license, I would opine that a world filled with attractions should seem more hospitable than one filled with repulsions.

Fig. 8.15 *Opposites can come together under the right circumstances.*

The like-likes-like mechanism forms the basis of the next chapter, which deals with Brownian motion: the jittery dance of suspended particles.

My interest in the Brownian phenomenon was triggered by an apparent paradox: Particles suspended in water ordinarily dance about. However, once a particle assumes membership in a colloid crystal, its motion practically ceases. The particle settles quietly into place at some distance from each of its near-stationary neighbors. Not only did this transition from motion to no motion seem curious, but so did the jittery dance itself: the particles seem possessed of a kind of internal energy, as though they were practically alive.

The issue continued to haunt me: why did particles seem so alive in suspension, while practically dying when recruited into an organized array? I could not stop thinking about it. Eventually, those thoughts gelled into the simple new understanding of Brownian motion detailed in the next chapter.

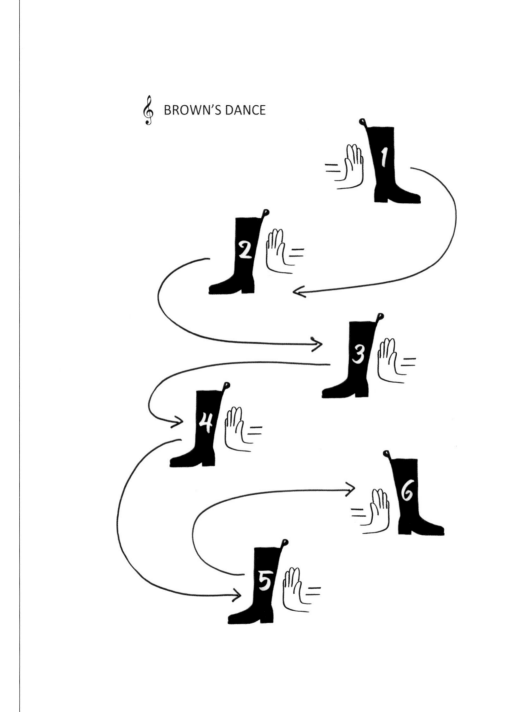

BROWN'S DANCE

9 Brown's Dance: Energy-Driven Movements

E arly in the nineteenth century, the Scottish botanist Robert Brown (**Fig. 9.1**) noticed something odd. After dropping some grains of pollen into a container of water, Brown observed a kind of jittery motion: instead of sitting lifelessly in the water, the grains danced around endlessly. He soon found that this motion characterized not only pollen, but also spores, dust, and even tiny window-glass fragments.

This jittery dance became known as Brownian motion, even though reports of similar motion had come half a century prior. Brown's main contribution was to demonstrate that not just biological entities exhibited this seemingly self-animated motion but also nonbiological entities. In other words, Brownian motion was a universal feature of nature.

What fuels these Brownian "dancers"? Why should inert particles move endlessly to and fro when "perpetual motion" is supposedly impossible? A common understanding of this peculiar motion has long prevailed, but that understanding has not considered the pool of radiant energy absorbed from the environment. We ask in this chapter whether that absorbed energy may bring a new (and perhaps simpler) understanding of this motion's origin.

Fig. 9.1 *Robert Brown (1773 – 1858).*

Brownian Motion According to Einstein

Brown's observations initially confounded physicists. Generating motion requires energy, but the source of that driving energy had remained unclear. The energy coming from the surroundings seemed an unlikely culprit, because scientists presumed that a beaker of water sitting on a table for some time was in equilibrium with the environment; the water did not appear to acquire any driving energy. Physicists scratched their collective heads, but no satisfying answer emerged.

Then came Einstein. In the seminal year of 1905, Einstein produced groundbreaking works on three fronts: special relativity, the photoelectric effect, and Brownian motion. He was having a good year. Einstein's explanation for Brownian motion did not require a constant input of driving energy. It rested instead on an ever-present internal energy that was reflected as temperature.

Einstein considered Brownian motion as arising largely from two phenomena: osmosis and friction. Osmosis is the phenomenon in which water moves toward solutes or particles; the concentration of water always wants to even out. Einstein took this inherent drive of water molecules to move as the motion's generator.

To appreciate how this might happen, imagine some particles suspended in water. An individual particle represents a location with no water. Because of the osmotic drive, water molecules always want to move toward that space. Occasionally, those water molecules will bump into the particle, provoking some movement.

Any such particle movement needs to overcome friction. Einstein recognized that. To deal with this viscous resistance, he applied the standard friction equation, known as Stokes' law of friction. He set the driving force of osmosis equal to the resistive force of friction and thereby articulated what has become the modern understanding of Brownian motion.

This simple sketch understates the sophistication of Einstein's analysis. Einstein dealt not only with the origin of the movement but also with its nature. He likened the movements of water molecules to the movements of gas molecules. Gas molecules, according to a kinetic theory well articulated by that time, bounce around randomly; their kinetic movements are inferred from their temperature. Temperature was considered as a measure of movement.

Thus, Einstein sought to extend that gas theory to liquids. By considering liquid molecules analogous to gas molecules, he could envision the water molecules jittering about randomly, occasionally striking suspended particles and pushing those particles to and fro (**Fig. 9.2**).

Of course, single hits will not produce very much motion. The mass of a water molecule is roughly 10,000,000,000 times smaller than that of the micron-sized particle to be moved. So its punch will be feeble. My friend Emilio Del Guidice likens it to the crash of a mosquito onto the

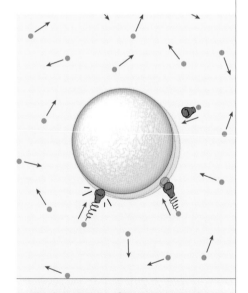

Fig. 9.2 *Origin of Brownian motion according to Einstein. Water molecules continually bombard the particle with mechanical energy. That energy produces movement.*

windshield of a trailer truck — the truck will not seriously deviate from its course. To move the truck, you'd need a lot of crashing mosquitoes.

At any rate, by using the kinetic theory of gases, Einstein arrived at an equation describing Brownian dynamics (see box, *page 156*). That equation predicts the magnitude of particle displacement (actually, the mean square particle displacement) over time.

With this formal theory, Einstein described what might happen to a particle suspended in a bath of water molecules. The water molecules, he opined, should undergo random gas-like motions. Sometimes those molecules would hit the particle. Because the hits would come at random instants, the particle should suffer random displacements, which might resemble the walk of a drunken sailor (**Fig. 9.3**).

Einstein's theory led to the concept of "thermal" motion. According to his formulation (see box), a particle's Brownian excursions in a liquid should depend on temperature. As the temperature increases, the excursions should increase concomitantly. On the basis of this temperature dependence, Einstein could refer to this motion as "heat motion" or "thermal motion." Physicists have come to take the temperature as a reflection of this motion: atoms and molecules dancing about with an intensity measured in terms of temperature.

While scientists broadly accept Einstein's analysis, acceptance was by no means the case initially. In his historical review, Brush[1] provides a readable account of the early resistance to Einstein's theory. Physicists expressed skepticism over what they considered questionable theoretical leaps. Stokes' law, for example, had been advanced for describing friction in macroscopic systems, such as swinging pendulums, but Einstein assumed that the law could also apply to the microscopic entities involved in Brownian motion. In addition, some physicists thought that the effects of bumping water molecules (like crashing mosquitoes) would lack enough power to explain the observed particle movements unless the collisions were coordinated. Other physicists raised concerns about mutually exclusive assumptions: osmotic theory implied that the water molecules should smash into the surface of the particle and then bounce off, but Stokes' law implied that those molecules had to remain adjacent to the particle in order to create friction. Those issues troubled physicists of the time.

Fig. 9.3 *Brownian motion, described as the walk of a drunken sailor.*

Brush goes on to describe another troublesome issue: Einstein's general approach. That approach rested on an abstract form of statistical mechanics, whose basis some physicists found difficult to follow.

To compound these theoretical difficulties, certain experimental observations contradicted Einstein's predictions. Contemporaries such as Svedberg and Henri found displacements four to seven times higher than those predicted from Einstein's formulation. These scientists were not shy about trumpeting those disagreements.

Nevertheless, as Einstein's stature grew, resistance to his theory gradually melted away; within several decades, Einstein's theoretical formulation became universally accepted. By now, thermal motion has come to be considered one of the most fundamental laws of nature. Every atom, molecule, and particle is thought to undergo this perpetual thermal dance; and out of such dance steps has come our contemporary understanding of condensed matter physics. It would not be an overstatement to suggest that the thermal motion concept has come to seem practically as fundamental as Newton's laws of motion, or the atomic theory of matter.

Concerning Issues

With the early reservations of Einstein's contemporaries now long forgotten, you might have thought that all issues Brownian are well settled. However, that is not so. The theory fails to match modern experimental observations in at least three scenarios: (*i*) when salt is added to the water; (*ii*) when particle concentrations are relatively high; and (*iii*) when a light is turned on. Let me briefly elaborate.

Regarding (*i*), Brownian excursions were measured in pure water and in water containing various amounts of salt.[2] Adding salt increased the excursions; i.e., it intensified the jitters. Einstein's analysis offers no obvious explanation for why the presence of salt might cause water molecules to bash into particles more frequently or more energetically.

Nor does Einstein's analysis anticipate an influence of particle concentration (*ii*). At high concentrations the motions of neighboring particles often become cooperative: when one particle moves, other particles nearby often move in the same direction.[3] Such synchronous behavior also shows up in colloid crystals[4] and, at high particle concentrations with added salt.[5] Synchronization creates a problem for the classical theory, which anticipates random motions. No coordination should appear.

A possible escape from this difficulty might involve claiming that the classical theory doesn't apply in high concentration situations: if the particle concentration grows high enough to squeeze out all of the water, then no water molecules would be on hand to hit the particles, so Einstein's formulation should no longer apply. However, particle concentrations rarely grew so high in the studies cited just above: the spacing between particles was typically on the micrometer scale, equal to a lineup of thousands of water molecules; hence, plenty of bouncy molecules should have been present to do the job. The conflict between theory and observation remains an issue.

Synchrony is not the only type of observed nonrandom behavior. At high particle concentrations, researchers detected a "hopping" motion.[6,7] A particle would jitter for some time within an almost stationary locus; then it would hop to a new locus, where it would once again suffer similar jittery displacements. It was as though particles jump from one cage into another. This caged pattern constitutes yet another apparent deviation from the theory's expectation of direct proportionality between the mean square distance traveled and time.

The effects of light (*iii*) bring an additional conflict. Although century-old studies by Gouy had denied any effects of light on Brownian excursions, more modern instruments have revealed profound and reproducible effects. Adding even modest amounts of light diminish particle excursions (**Fig. 9.4**). The reduction depends on intensity and wavelength and can easily amount to 50 percent.[8]

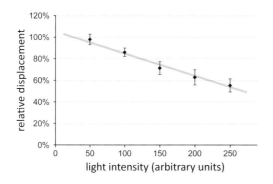

Fig. 9.4 *Microsphere displacements measured over a fixed period of time at different incident light intensities. Higher light intensities diminish displacements.*

These light effects could be brushed aside if they were mediated indirectly, through light-induced temperature increase; however, the temperature increase in the above-cited experiments amounted to less than 1° C. Actually, temperature fails as a theoretical explanation: Einstein's formulation anticipates that any temperature increase should *enhance* particle excursions, but light induced *smaller* excursions. So, that expedient cannot suffice in any case to rescue Einstein's explanation.

The light-mediated effect has no obvious explanation within the classical framework. On the other hand, the effect is striking enough to demand explanation. We will tackle this issue shortly.

Apart from the three modern categories of conflict just considered, a fourth issue arises from the uncertain nature of the driving force. Einstein suggested osmosis as the driver of Brownian motion. He did so because osmotic movement, in his day, was presumed to be a fundamental feature of nature, on the same level as, say, the attraction between positive and negative charges. However, the osmotic mechanism has since become more uncertain, and has remained under debate. Chapter 11 presents evidence that the osmotic drive is a consequence of charge separation, and not the fundamental feature of nature that the classical theory posits.

In sum, while Einstein's theory of Brownian motion has broad acceptance and fits some experimental observations, it fails to fit others, not only from the past but also the present. These failures imply that the theory is at best incomplete. More than incomplete, I argue next that the theory is inadequate: it fails to take account of the drive supplied by absorbed incident energy, which may influence, or even cause, the observed Brownian motions.

Nonequilibrium: Another Brownian Issue

Einstein's formulation hinges on the presumption of a system at equilibrium. So long as the ambient temperature remains steady, the system should neither gain nor lose energy. A pot of warm water certainly loses energy to the environment, and a glass of cold water may gain energy; but a covered vessel of water sitting in a room for some time should qualify as neither losing nor gaining energy. So the classical theory presumes.

However, Chapter 7 offered evidence that room-temperature water continually absorbs electromagnetic energy from the environment. That absorbed energy builds order and separates charge — creating potential energy that can subsequently produce many kinds of work. By mediating this conversion of energy, water functions like all common working engines: out of equilibrium.

Out-of-equilibrium behavior characterizes even one of life's most primitive water-based mechanisms: photosynthesis (**Fig. 9.5**). In photosynthesis, incident photons split water, the products of which go on to drive metabolism, growth, flows, *etc*. Absorbed light energy continually drives work output — which means that the system lies out of equilibrium.

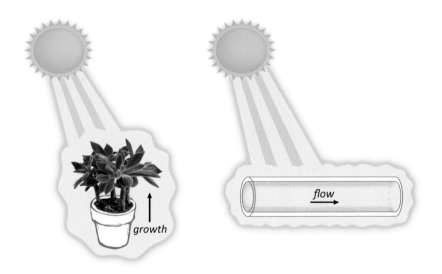

Fig. 9.5 *Radiant energy produces work in plants and in water; the two processes are conceptually similar. By processing input energy, both operate out of equilibrium.*

Hence, the presumption of equilibrium in classical theory conflicts with the evidence that *water is out of equilibrium*. Both water and plants (the latter consisting mainly of water) operate out of equilibrium: input energy continuously powers work in much the same way as fuel continuously powers your car. If water lies out of equilibrium, then any understanding of Brownian motion built on the presumption of equilibrium must be considered suspect.

The issue is serious. Particles continuously move and do work, but in the classical view, no outside energy is needed to drive that work. Internal energy suffices — indefinitely. How the work of movement could persist without any external assistance presents a challenge to the classical understanding of Brownian motion.

An Alternative Driver of Brownian Motion?

Given these theoretical and experimental challenges, it seems worthwhile to take a step back and reflect. A century has passed since Einstein proposed internal heat as the driver of Brownian motion. Stature notwithstanding, can we be certain that Einstein's proposal is necessarily valid?

Let me propose an alternative Brownian motion driver: incident electromagnetic energy. If absorbed electromagnetic energy drives other water-based transduction processes (Chapter 7), then might that energy also drive Brownian motion?

A hint that this hypothesis may hold promise can be glimpsed by recalling water flowing through hydrophilic tubes. Think of that flow as the sum of many directed Brownian motions. From that perspective, the tube merely organizes those motions into a collective flow. Since electromagnetic energy drives the tubular flow, electromagnetic energy logically drives the individual motions that collectively make up the flow (**Fig. 9.6**).

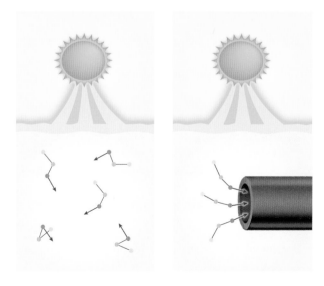

Fig. 9.6 *Incident energy generates motions, which can be random or coordinated, in the absence* (left), *or the presence* (right) *of a director.*

An external Brownian driver also makes sense from an energy-flow perspective. Water continuously absorbs electromagnetic energy — which needs to be expended. We might therefore view Brownian work as a kind of relief valve, a way to expend all of that absorbed energy. Put another way, Brownian motion may be a natural reflection of water's continuous absorption of electromagnetic energy.

The idea of an electromagnetic driver of Brownian motion may seem radical, but it has precedent. Several 19th century physicists thought that electromagnetic energy might drive the motion indirectly by inducing heat. Later, Max Planck, the father of quantum mechanics, thought along similar lines: he entertained the notion that electromagnetic interactions could drive random molecular motions. Planck finally moved in other directions when that line of reasoning failed to yield the rich dividends he had envisioned. Nonetheless, the idea of an electromagnetic driver had remained high on Planck's agenda for two decades. The notion is not so radical and is not without rationale.

The Force Driving Brownian Motions

While the considerations above may help put Brownian energetics into a plausible framework, they tell us nothing of the immediate driver. What force might push the particles to and fro? If we accept provisionally that electromagnetic energy might drive these motions, then how does this energy translate into movement?

Occam's razor suggests that we consider some EZ involvement. If electromagnetic energy builds exclusion zones, then some feature of the EZ ought to enter into the equation. The EZ has two principal attributes: molecular order and charge separation. Both attributes seem worth considering. Charge, in particular, can generate substantial forces, which could conceivably drive the Brownian jitterbug.

To see how charge might drive a particle, imagine a single microsphere suspended in water. The microsphere contains a shell-like EZ. If energy were uniformly incident from all directions, then that EZ should be uniform. In practical circumstances, however, absolute uniformity of incident energy is unachievable; and with any nonuniformity, we expect a nonuniform shell with a correspondingly nonuniform charge distribution; see **Figure 9.7**.

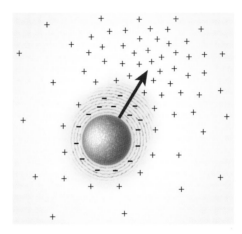

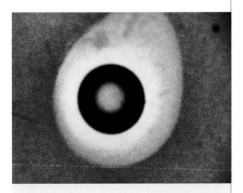

Fig. 9.7 *More intense incident energy (from upper right) should yield an asymmetric charge distribution. This asymmetry gives rise to a net electrostatic force, moving the object and its EZ in the direction of maximum incident energy.*

Fig. 9.8 *Asymmetric exclusion zone around a gel bead, resulting from incident light coming from top right.*

In this figure, which way would the suspended microsphere want to move? The movable entity is not the microsphere alone; it is the microsphere plus its clinging EZ. That blob could move in any direction in theory, but being negatively charged, it will inevitably move in the direction of highest positivity. In the figure, that is toward the upper right.

We tried to model the EZ asymmetry of **Figure 9.7** and got a positive result. **Figure 9.8** shows a gel bead (approximately 0.5 mm in diameter) sitting on the floor of a chamber. The exclusion zone ordinarily envelops the bead uniformly. However, when we shone extra light from one direction, the EZ grew larger in the direction of the light. Hence the asymmetry implicit in **Figure 9.7** is physically realizable.

Next, we tested whether a negative EZ moves toward positive charge, as **Figure 9.7** implies. To do so, we used a different setup. We positioned an assembly, consisting mainly of a long ribbon, horizontally in an experimental chamber so that the ribbon's edges lay at top and bottom. The ribbon was cantilevered from one end (**Fig. 9.9,** *panel a*). Near the opposite, free end, we bonded a Nafion tube to one face of the ribbon.

When we exposed the ribbon-Nafion assembly to water, the EZ grew around the Nafion, releasing protons in the usual way (*panel b*). However, the ribbon was large enough to block proton diffusion, so the protons remained on one side of the ribbon. As those protons spread, the lever bent impressively toward them (see *panel c*).[9] This confirmed the expected: negatively charged EZs move toward positively charged protons — even toward the protons that the EZs themselves create. This affirms the principle illustrated in **Figure 9.7**.

Thus, we expect that particles such as those illustrated in **Figures 9.7** and **9.8** will always move in the direction of highest local positivity — toward the upper right, in those figures. All that's needed is local charge asymmetry, an almost inevitable condition in an array of particles subject to nonuniform incident energy. The movement toward positive charge may be thought of as *the basic unit of Brownian excursion*.

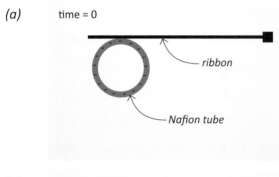

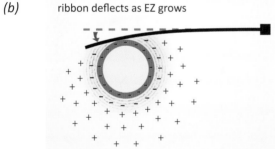

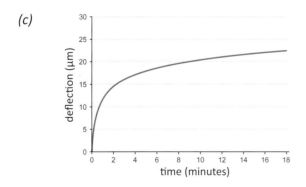

Fig. 9.9 *EZ deflection toward protons.* (a) *Nafion tube mounted on ribbon face;* (b) *ribbon deflecting toward protons;* (c) *measured time course of deflection.*

Do Particles Really Move Toward Light?

If this attraction-to-charge principle genuinely operates, then we should expect consequences. One consequence follows from the effect of incident light. Incident light builds EZs and separates charge around particles. If the light comes mainly from one direction, then charge will separate more in that direction than others. Suspended particles should therefore move toward the incident light.

I'll show you four examples that confirm this behavior.

• Attraction to light was evident, though subtle, in the previous chapter. I mentioned that extra light draws colloidal particles closer together. It does so by separating the charges that mediate the attraction. Thus, regions receiving more light will experience more attraction and more condensation. That condensation continuously draws in additional microspheres — which is tantamount to saying that the microspheres are drawn toward light.

• A second piece of evidence for attraction to light comes from experiments that restricted the beam of incident light. We passed a beam of light through a hole and into a uniform suspension of microspheres. The microspheres moved toward the region receiving the light, eventually concentrating in the narrowly illuminated zone (**Fig. 9.10**).

Fig. 9.10 *Light-induced attraction of microspheres. Light shines through a hole in an aluminum mask. After some time, microspheres in the chamber beneath the mask gather at the center of the hole.*

microspheres

Bacteria do much the same (**Fig. 9.11**). They move toward near-infrared light in the same way that the microspheres above moved toward the restricted light beam. The bacterial movement is hypothesized to arise from an infrared sensor lodged inside the cell.[10] While that might be the case, the movement toward light so closely resembles that of the microspheres that one may rightly ask whether the physical mechanism under consideration might be at play.

• A third example: do you recall the vertically oriented, microsphere-free cylinder that formed near the middle of the beaker? I introduced this anomaly in Chapter 1. Initially, the microspheres were distributed uniformly throughout the beaker of water. Eventually they migrated toward the beaker's periphery, leaving a vertically oriented cylinder devoid of microspheres (**Fig. 9.12**). What drew those microspheres, we found, was the light impinging on the beaker from all directions. That incident light drew the microspheres toward the beaker's periphery, leaving the middle empty.

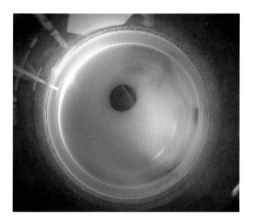

Once the cylinder formed, we could further investigate the effect of light. Shining light from one side quickly drew the microspheres toward that side; the shift displaced the cylinder progressively toward the unilluminated side, where it eventually collapsed into nothingness. All of this happened within a minute or two.[11]

We have seen similar light-mediated displacements in other configurations. For example, we attached a gel disc to one end of a tall, cylindrical chamber, which we then filled with water and microspheres;

Fig. 9.11 *Bacterial cells and particles move toward the point of highest light intensity.*

Fig. 9.12 *Microsphere-free zone in a beaker (viewed from above). Running from top to bottom, the clear zone appears near the center of the aqueous microsphere suspension.*

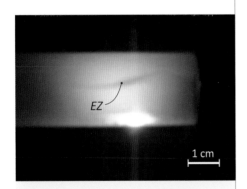

Fig. 9.13 *Light-induced deflection of microsphere-free zone. This EZ projects from a nucleating surface positioned off to the left. The light beam projects from bottom to top. Microspheres drawn toward the light cause the EZ to shift oppositely.*

we then laid the cylinder on its side and watched the exclusion zone develop. The EZ changed shape — it started as a disc (reflecting the shape of the nucleating gel), but narrowed down in a roughly conical shape, finally taking on a more pole-like shape as it continued to grow along the cylinder. We often examined those pole-like EZs with a flashlight or laser beam. Each time we turned on the light, the EZ would shift away from the light source (**Fig. 9.13**). The shift seemed to occur as a byproduct of the surrounding microspheres' drawing toward the light source, as described above.

• A fourth example relates to a phenomenon we observe in ordinary microsphere suspensions: sedimentation. Microspheres generally settle to the bottom of the chamber after some time, forming a sedimentary layer. We found that shining a light from above retarded the sedimentation, whereas shining a light from below accelerated it. Again, the light drew the microspheres.

These multiple examples leave little doubt that suspended particles move toward light. Thus, we are not whistling in the dark: we have confirmed the reality of the light-mediated force hypothesized to drive Brownian movement.

Ensemble Dynamics

The remaining issue is just how those light-driven displacements create the seemingly random motions characteristic of Brownian phenomena. Up to now, we have focused on the single particle: incident light creates an asymmetric charge distribution around the particle, which draws the particle toward the highest positive charge and hence toward the highest-intensity light.

When multiple particles populate the water, the scenario grows more complex (**Figure 9.14**). One microsphere's EZ generates positive charges. Those positive charges attract neighboring microspheres, which move in response. They move along with their EZs and clinging hydronium ions. Those movements alter the regional charge distribution. They may also block or unblock light pathways to other microspheres, and so on. The dynamics become so complex as to seem random.

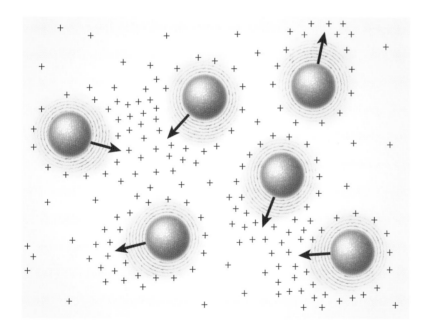

Fig. 9.14 *Charge distributions surrounding microspheres in suspension. Arrows denote anticipated directions of movement of negative particles toward positive-charge maxima. Directions will continually change as particles move.*

To complicate matters further, positive ions liberated from one EZ may affect the size of another microsphere's EZ: we found that EZ size depends on local positive ion concentrations. Hence, the multiple particle scenario becomes complex enough to make deterministic predictions of particle movements practically impossible.

On the other hand, if we are on target, then local dynamics might be somewhat predictable. Since the movement of one particle should influence the movement of neighboring particles, displacements of neighbors should be loosely coupled. This should be most evident at high particle concentrations, when the charge of one particle can more strongly influence the position of another. Experiments have confirmed this: as already described, coupling does occur, and it occurs most clearly at high particle concentrations. This conforms to our expectation.

Merits of the Light-Driven Mechanism

Ultimately, we must ask whether the proposed Brownian mechanism has better explanatory power than the classical mechanism.

The classical formulation works (with the important exceptions noted); otherwise, it would not have survived as long as it has. That formulation contains three variables (see box): fluid viscosity, temperature, and particle size. From the way the equation arranges those variables, Brownian excursions should diminish if you do any of the following: increase fluid viscosity; increase particle size; or reduce fluid temperature. Experimentation has confirmed all of these expectations. Hence, the classical model suffices for explaining those basic variables.

On the other hand, the new formulation also works. To wit:

- *Viscosity:* As you'd note if you immersed a bumblebee into a jar of honey, viscosity dampens the jitters. This dampening should apply irrespective of the nature of the driver producing the jitters. The principle fits both new and old formulations alike.

Einstein's Theory of Brownian Motion

Einstein's original equation described the diffusion constant, D, of a particle in a fluid:

$$D = \frac{k_b T}{6\pi\eta a}$$

where k_b is Boltzmann's constant, T is the absolute temperature, η is the fluid viscosity, and a is the particle radius.

From the value of D, one can calculate the displacement, x, over time as:

$$\overline{x^2} = 2Dt$$

where $\overline{x^2}$ is the mean square distance traveled, and t is time. This equation lets us predict how far a particle should move in a given amount of time.

• *Particle size:* Larger particles should jitter less than smaller particles because the larger particles need to push aside more water molecules in order to move. This expectation follows from almost any theoretical formulation, and it applies here as well.

• *Temperature:* As experimentally confirmed over temperatures ranging from 0 to 30 °C, reducing the water temperature increases EZ size. Increasing EZ size expands the microsphere's effective bulk. That creates a larger mass, which means that the particle cannot move as far in a given window of time. Hence, the temperature dependence fits the new formulation, just as it fits the old.

Beyond accommodating those three basic expectations, the EZ mechanism also explains a curious phenomenon: microspheres almost cease to move if they join a colloid crystal array. A microsphere lying just outside the array will jitter with the usual dynamics; but, if it joins the ordered array, it will slow nearly to a standstill (**Fig. 9.15**). This does not result from any physical constraint imposed by neighboring microspheres, for the usual several-micrometer span between microspheres means that many water molecules lie in between. Microspheres don't stop moving because they butt up against one another.

The reason those microspheres come to a virtual halt lies in like-likes-like constraints. The colloid crystal achieves its stability because strong attractive and repulsive forces lie firmly in balance; that tight balance confers stability. It ensures that microspheres lying within the array become relatively immune to charge fluctuations arising outside the array and thereby suffer fewer Brownian fluctuations. Thus, the newer formulation plausibly explains the otherwise paradoxical behavior illustrated in **Figure 9.15**.

Another feature that the proffered mechanism needs to explain is the presence of Brownian motion in liquids other than water. Exclusion zones appear in many polar solvents, and those EZs separate charge.[13] The new formulation does not fail us; it anticipates Brownian motions in those liquids as well as in water.

The EZ mechanism must also explain the motion's endlessness. You can put the beaker of water and microspheres away for a day — or a year — and (provided the microspheres don't sediment to the bottom)

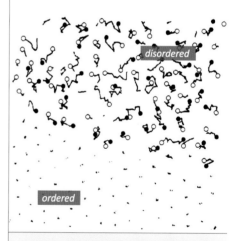

Fig. 9.15 *Traces of particle movement over a period of time, both within the ordered region (lower left) and outside the ordered region (upper right). From Dosho et al.[12]*

the motions will continue unchanged. Those particles just keep jittering. The jitter is endless because the driving energy is endless: so long as the liquid continues to absorb electromagnetic energy, that absorbed energy will continue to drive the Brownian motions.

Why does dust undergo Brownian motion?

Dust comprises mostly flakes of skin and hair. Both are negatively charged; hence, they repel. Repulsion builds further as the dust passes through the air: the dust acquires charge in the same way as your hair acquires charge (and fluff) from a hair dryer — the triboelectric effect. The faster the relative movement of the air, the higher will be the negative charge and, hence, the repulsion.

The atmosphere, on the other hand, contains positive charge. The air's known positive charge can neutralize the dust's negative charge. The process takes a while because airborne charges need time to gather round the particle. Hence, slowly moving particles get readily neutralized, while rapidly moving ones do not. So the particle's speed counts. A particle's net charge will therefore be highly dynamic, leading to the seemingly erratic Brownian air dance.

You might wonder why the moving dust particles seem to float as they dance. Denser than air, those particles should steadily descend toward the earth; yet they float. At play here is the earth's net negative charge — an attribute well established but little recognized. The earth's negative charge repels the dust's negative charge; hence, the particles stay afloat. They continue their endless dance, rarely settling, and never touching one another in their mutual repulsion.

Implications

Lucretius's classic scientific poem *On the Nature of Things (ca. 60 BCE)* provides a memorable description of Brownian motion — albeit in dust particles:

"Observe what happens when sunbeams are admitted into a building and shed light on its shadowy places. You will see a multitude of tiny particles mingling in a multitude of ways … . [T]heir dancing is an actual indication of underlying movements of matter that are hidden from our sight … . It originates with the atoms, which move of themselves [i.e., spontaneously]. Then those small compound bodies that are least removed from the impetus of the atoms are set in motion by the impact of their invisible blows and in turn cannon against slightly larger bodies. So the movement mounts up from the atoms and gradually emerges to the level of our senses, so that those bodies are in motion that we see in sunbeams, moved by blows that remain invisible."

Lucretius presciently describes the contemporary view of the origin of Brownian motion. Every atom, every molecule, every particle, and every larger entity suffers random displacements. The smaller ones hit the bigger ones, driving them into motion. This action was nicely described two millennia ago. Until the contribution of Einstein, however, nobody understood the origin of those displacements. Einstein concluded that the heat contained *within* the system was the driving force. That heat generated motion, which produced heat, which generated motion, *etc.*, a process continuing endlessly without needing any outside assist.

In Einstein's day, however, scientists could not conceive that a simple aqueous suspension could absorb energy from *outside* the system and put that energy to use. Even though this happens in plants all the time, nobody could imagine it occurring in nonliving systems, such as beakers of water. Yet the evidence presented above shows that it can and does happen: absorbed energy gets continuously absorbed and put to good use, one of those "uses" being the driving of Brownian motions.

If this new explanation of Brownian motion proves valid, many physical phenomena might need rethinking. An important one is "thermal motion." Thermal motion is the common terminology for Brownian motion. Until now, it has been thought that thermal motions of atoms and molecules are driven by *internal* energy. If, instead, *external energy* drives these motions, then a very different paradigm emerges with quite different consequences.

One important distinction between the two formulations lies in the influence of neighbors (**Fig. 9.16**). In Einstein's formulation, particle movements depend only on the blows of abutting water molecules; any particles lying beyond those molecules don't count much. In the EZ formulation, the opposite is true: particles lying at some distance generate charges, and those fluctuating charges may influence the motion of the particle in question. Those effects can be long-range. In this sense, the two proposals differ fundamentally, over and above their mechanical versus electrical origins.

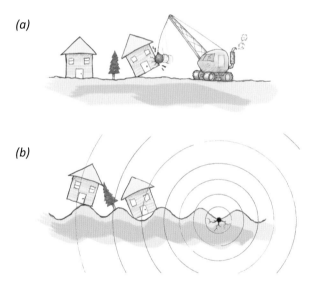

(a)

(b)

Fig. 9.16 (a) *Einstein's formulation emphasizes local influence.* (b) *The EZ formulation implies long-range influence.*

As a consequence of these distinctions, phenomena that seem anomalous within the Einsteinian paradigm become less so within the EZ paradigm. I already mentioned the coupling of nearby particle displacements and the virtual absence of particle displacements within the colloid crystal. Neither phenomenon fits in the classical paradigm, but the EZ paradigm accommodates them both. The salt-induced jitter also finds no easy explanation in the classical paradigm. The new paradigm may make this a simple matter of size: salt diminishes EZ size,[14] which effectively diminishes particle size, allowing the particles to dance more actively. Thus, at least some phenomena difficult to reconcile with the classical paradigm find natural explanations within the EZ paradigm.

Whether the EZ formulation explains all features of Brownian motion remains unclear, but I believe it has a chance because it contains a fresh new feature: external energy input. That feature may on the one hand require a rethinking of the relation between EZ growth and conventional entropy, while on the other hand, may unlock many unresolved mysteries of Brownian dynamics.

Before moving forward, however, I feel the need to resolve a few issues. Did the concept of heat-induced motion seem clear as you read through this chapter? Perhaps it did. When I first learned that internal heat drove Brownian motions, I must admit to a sense of confusion: I couldn't understand how heat drives motion, although I understood that physicists consider heat and motion practically synonymous. The association was familiar, but the underlying mechanism seemed vague.

Heat and temperature are terms we use freely, but I came to find that their meanings were less straightforward than generally presumed. Understanding those concepts in an intuitively satisfying way requires fresh consideration, and the next chapter attempts to meet that challenge.

Summary

According to conventional views, Brownian (thermal) motion arises from the molecular kinetic energy that we commonly express in terms of temperature. This energy is thought to drive particles endlessly to and fro in a random (Brownian) fashion. Although that theory of Brownian motion has become universally taught, a surprising number of experimental observations do not fit.

The alternative hypothesis put forth here suggests that incident radiant energy drives Brownian motions. The absorbed energy builds exclusion zones around the particles and thereby separates charge. The separated charges generate forces that drive particle movements.

Its consistency with experimental evidence lends confidence in this alternative model. The model is also intuitively straightforward: energy input drives energy output. Hence, this rather simple model may finally begin to resolve the many paradoxes surrounding Brown's dance. We may finally understand why those particles jitter endlessly.

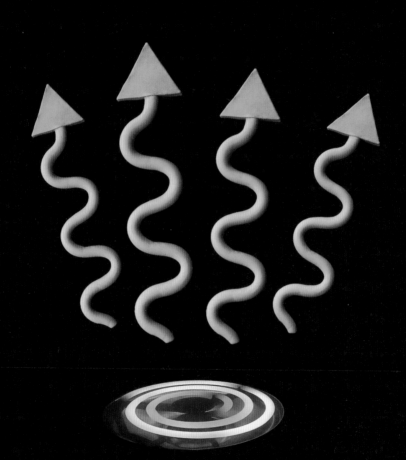

10 Heat and Temperature: Throwing New Light on Thermal Darkness

While lunching at a charming island restaurant near Seattle, a colleague's passing comment caught my interest. The comment referred to temperature. My colleague said that if you vigorously swished a container of water around to form a vortex, the water would cool. "No way!" I exclaimed. Swishing generates friction. Everyone knows that friction induces heating, not cooling. He had to be wrong.

My colleague turned out to be right after all. I followed up our lunchtime conversation by asking one of my students to repeat the experiment. He indicated that he'd already vortexed laboratory water many times and confirmed that it got cooler. I'd heard the same from a New Zealand colleague, who vowed to determine how much cooler the water could get but never made it beyond a drop of 4 °C.

Vortices are natural phenomena seen in rivers and streams, as well as in bathtub drains and flushing toilets. A popular demonstration is illustrated in **Figure 10.1**.

Why does vortexed water cool?

You'd think we could grasp the essentials by being properly rigorous in dealing with heat and temperature. Rigor usually helps. However, as the last chapter showed, even proper rigor applied to a question-able foundation cannot necessarily produce a sound result. Brownian motion was long supposed to be driven by heat, yet rigorous treatment by renowned physicists never managed to produce a fully satisfying understanding. Some elements of that understanding remained amiss — some of them evidently basic.

Those key elements involved heat and temperature. The terms "heat" and "temperature" (along with "entropy") seem central to prac-tically every energetic consideration. Yet those terms are surprisingly vague, as we shall see. Using those terms in casual conversation may be fine; but building understanding on vaguely defined terms can prove

Fig. 10.1 *Common demonstration of water vortex.*

hazardous. Doing so can lead you to think that something ought to get hotter when in fact it becomes cooler.

Because of this risk, we will avoid using vague terms and stick to more rigorously defined ones. One such term, radiant energy, relates to temperature and heat, but it has the advantage of carrying a unique definition. On the other hand, some find the language of radiant energy unfamiliar; I hope you will bear with me through the short "tutorial" that begins this chapter. Small pains may yield big gains.

The Origin of Radiant Energy

Radiant energy is electromagnetic energy. It may encompass a broad range of wavelengths, each segment of the spectrum exhibiting different characteristics (**Fig. 10.2**): light waves can be seen; microwaves can cook your food; radio waves can help us communicate; X-rays can produce images; and infrared waves can keep us warm. These features seem so different from one another that we can easily forget that the waves all belong to a single electromagnetic spectrum.

Fig. 10.2 *Various sources of radiant energy. Radiant energy includes a broad array of wavelengths.*

To understand the workings of radiant energy, you need to under-stand how those electromagnetic waves originate: *always by the move-ment of charges*. **Figure 10.3** schematizes this concept. Envision a static charge positioned somewhere in space (*left panel*). If situated close enough, you (or your detector) may sense that charge. If the charge moves, then your relationship to the charge changes (*middle*). Sens-ing that change may take some time, depending on how quickly the information can propagate through the intervening medium. Similarly, back-and-forth movement of charge will create a back-and-forth oscil-lation (*right*), which again you will sense after a brief interval. You are now sensing a propagating electromagnetic wave.

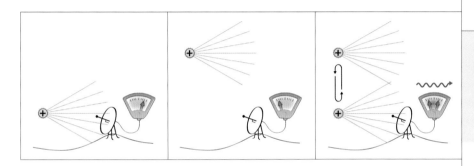

Fig. 10.3 *Electromagnetic wave gen-eration (simplified). Back-and-forth movement of the charge creates an oscillating electric field, which a sensor can pick up.*

Any charge oscillation can serve as the generator of a wave. The oscillat-ing charge can come from an electron, a proton, a nucleus, or even some larger charge-bearing entity. All qualify as generators. Also, the displacement can range from miniscule, as within an atom, to colossal, as in a large radiating antenna. The wave-generating process, however, is always the same: charges oscillating back and forth.

Interaction of Waves with Materials

Now consider what happens when an electromagnetic wave passes through a material. All materials contain charges. Waves exert forces on those charges; hence, the incoming wave will push or pull on what-ever material charges they encounter. Those charges will respond by moving. If the incoming wave is periodic, then the material charges should oscillate at the same periodicity. Effectively, a charge oscillation induces another charge oscillation. Each receiver becomes the next generator. The process continues.

Just how it continues depends on the medium through which the wave passes. If the medium is the same throughout, then the wave's principal features will remain invariant, although the wave might attenuate. If the medium lacks uniformity, then the oscillation's character will vary as it passes through. For example, it might propagate more quickly through one region than another; quicker propagation means traveling a longer distance over the same time between peaks, which means it has a longer wavelength. So wavelengths can change as the wave travels through the material.

Finally, those waves emerge from the medium. The emerging waves may differ from the incident waves because of the changes just described. An incident wave of 10-μm wavelength may propagate through the medium, get absorbed, and then reradiate at, say, 5 μm or 20 μm, depending on the medium's character. Waves passing through complex media can reradiate at longer or shorter wavelengths (so-called Stokes and anti-Stokes shifts); see **Figure 10.4**.

Fluorescence provides an example of such a shift. Incident light of some wavelength temporarily shifts the material's electrons to higher energy levels; as those electrons drop back down, they radiate at a longer wavelength. Thus, incident blue light may cause the material to emit red light. The material is said to fluoresce red.

Spectral shifts also occur at IR wavelengths. A familiar example is your house. Sunlight strikes the outer walls, which absorb the incident energy. The walls then radiate some of that energy to the interior walls, which in turn radiate into the room — so you feel warm. The radiant wavelengths and amplitudes emerging from those inner walls may differ entirely from those of the incoming sunlight.

These examples illustrate a fundamental dynamic: radiation entering a system drives charge movement, which generates electromagnetic waves, which drive charge movement, *etc*. Finally, the waves exit the system, but only after having suffered multiple wavelength and amplitude shifts, and perhaps after having powered some work (Chapters 7 and 9). So, a material's radiant emission depends not only on what enters the system, but also on the character of the medium.

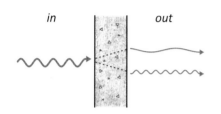

in   *out*

Fig. 10.4 *Electromagnetic radiation passing through a complex medium. The character of the radiation may change. Total output energy need not equal input energy if the medium stores energy and uses it to produce work.*

Radiant Emission from Water

How can the character of the medium help us understand what occurs in water?

The emissive character of a medium generally gets expressed by a shorthand term: "emissivity." Objects with higher emissivity radiate more energy than objects with lower emissivity; they seem more energized. If their radiant emission happens to lie within the IR wavelength range, then objects with higher emissivity will appear brighter in an IR camera than neighboring objects with lower emissivity.

For example, consider **Figure 10.5**. The standard visible light image of an office wall (*top*) shows only dull, rudimentary detail. The IR image (*bottom*) reveals more of the underlying structure. Some of the image's rich detail derives from emissivity differences.

A more telling infrared image, showing clouds, appears in **Figure 10.6**. Regions of those clouds generate substantial radiant energy. Conventional lines of thinking relate IR intensity directly to temperature (see temperature scale on the right side of the image). Thus, an expert steeped in current understandings of such phenomena might say that the cloud is hotter than the frigid winter sky around it. The wintry cloud also reads as hotter than the working smokestack beneath. Those interpretations make little sense. Something must be amiss with the temperature-based interpretational framework.

Fig. 10.5 *Interior wall in a room situated just off a corridor. Observed with visible light* (top) *and with IR light* (bottom).[1] *The IR image reveals details of the wall's internal structure.*

Fig. 10.6 *IR image, obtained in the wavelength range 9–12 μm. Ground temperature, approximately 0 °C. Smokestack and treetops are visible at bottom. On the temperature scale provided by the camera manufacturer, distant cloud regions register hotter than 15 °C despite a surrounding air temperature of minus 20 °C. Thus, the temperature scale paradoxically implies a sustained temperature gradient of 35 °C.*

The radiant energy framework, by contrast, implies that the cloud has a higher emissive character than its surroundings; the moving charges inside the cloud simply generate abundant infrared energy. The cloud may look "hotter," but really the cloud's charges are bouncing around more.

This illustration shows how blind reliance on familiar concepts can sometimes mislead. Surely the cloud in the frigid sky cannot act as an oven in a refrigerator. The misinterpretation stems from reliance on familiar terms such as heat and temperature. It raises big red flags over the use of those common terms.

Consider a third IR image, which you've already seen (reproduced here for convenience). **Figure 3.15** shows an IR image of the water near a Nafion surface. Since the respective EZ and bulk water zones had remained long juxtaposed by the time this image was taken, any physical difference between those zones should have equilibrated; even so, the exclusion zone remains darker. It emits less infrared energy. We could easily say the EZ has a lower "temperature" than the bulk water next to it, but that vernacular once again misleads.

Why should the EZ emit less radiant energy than bulk water? Think about charge movements in the respective zones. In the EZ, charges have become fixed in a lattice; they may jump from one lattice point to another, but for the most part, those charges remain in place. In the adjacent bulk water zone, the charges are free: they may move with some abandon. Since moving charges generate radiant energy, the bulk-water region should appear brighter. For that reason, the bulk water looks "hotter" than the EZ, but strictly speaking, its charges are simply moving around more actively.

So, brightness and darkness in IR images do not necessarily imply higher or lower *temperature*; more accurately, they reflect higher or lower intensities of charge movement. Referring to the brighter objects as having "higher temperature" may make sense for casual conversation, but for scientific discourse, a safer bet is to avoid "temperature" and "heat" altogether, sticking with more defensible terms like radiant energy.

That point constitutes this chapter's central take-home message: *radiant emission reflects the intensity of charge movement*. This is true for all wavelengths

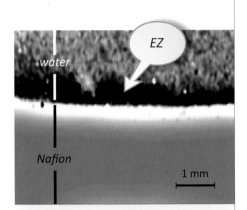

Fig. 3.15 *Infrared emission image of Nafion next to water. The sample was equilibrated at room temperature. Black band running horizontally across the middle of the image corresponds to the expected location of the exclusion zone.*

of the electromagnetic spectrum. If we adhere strictly to this fundamental linkage without succumbing to the temptation to invoke temperature or heat, we should not go far astray in our search for understanding.

What Exactly Are Temperature and Heat?

A word about the origin of those familiar terms seems in order. To understand why they confuse, one needs to know what they mean.

As for "heat," no single definition prevails. Heating often (but not always) gets described as the transfer of energy from one physical body to another, excluding any work done on that body. Thus, you can supply radiant energy to a body and label it as heating; but you can't raise a boulder to the top of a mountain and call it heating, for work doesn't count. In principle, the more radiant energy you supply, the more heat you add.

Now consider heating in relation to water. Like other materials, water absorbs, transforms, and then reradiates incident radiant energy. The most relevant wavelengths lie in the IR region of the spectrum, particularly 3 μm to 15 μm. We can find a reason for this range: the water molecule's charges reside at certain characteristic distances from one another, which means that whenever those charges oscillate, they prefer to do so at wavelengths linked to those characteristic spacings, namely 3 μm to 15 μm. Water will preferentially absorb and emit radiation at those infrared wavelengths.

Thus, we can appreciate why "infrared" and "heating" often come in practically the same breath when dealing with water. Water absorbs infrared; it therefore "heats up." Water also emits infrared, and therefore it "feels warm."

We need to bear in mind, however, that "infrared" and "heating" are not interchangeable. Water absorbs not just those IR wavelengths, but also many wavelengths across the electromagnetic spectrum. Even visible light can heat water if you provide enough of it. And the microwave energy in your oven can heat water quite efficiently. So, *heating does not uniquely equate to IR absorption*. Nor can we assert that heated water radiates at IR wavelengths alone. Indeed, water can emit energy even at visible wavelengths (Chapter 7).

Because of the vague relationship between radiant energy and heat, it is only with the greatest of caution that we dare use the term "heat" in trying to improve our understanding.

Now consider "temperature." When water "heats up" from IR or other absorbed energy, we say that the water's temperature increases. Again, we need to know: what exactly do we mean by temperature?

Unfortunately, no unique definition exists for temperature either. The definition depends on the field. Some definitions include the following: the degree or intensity of heat present; the ability of a substance to transfer heat energy to another substance; a measure of the average kinetic energy of the atoms or molecules of a substance; a reflection of particle motion arising out of translation, vibration, or excitation of an electron energy level; and, in a gas, the probability distribution of the energy of motion of gas particles.

Even temperature's well-known end points don't help bring clarity. Water is said to freeze at 0 °C and boil at 100 °C. You'd think that these benchmarks would constitute useful reference points for pinning down temperature's real meaning. But they do not, for pure water at standard pressure can freeze at temperatures much lower than 0 °C (especially in confined spaces), and it can vaporize at temperatures above, and sometimes below, 100 °C.[2] We might choose to relegate these variants to anomalies, but perhaps they indicate an inherent ambiguity in our understanding. Given its multiplicity of definitions, "temperature" seems just as ambiguous as "heat."

Adding to the problems of definitional ambiguities, we have a more fundamental problem for systems out of equilibrium — like water. Thermodynamicists warn that, for such systems, empirical measures may disagree on which of any two bodies is hotter. Translated into plain English, that means that any system out of equilibrium has no well-defined temperature. For water, this is a profound problem. Until we resolve it, *we simply cannot know the meanings of "temperature" and "heat."*

You can now understand why I refrain as much as possible from using those familiar terms. In everyday conversation, what can substitute for "the stove is hot!"? For scientific discussions, however, squishy definitions inevitably lead to squishy conclusions, and even erroneous ones. Brownian motion provides one example; vortexing, another. In both of those cases we are misled by the use of temperature as a foundational variable.

Real understanding, on the other hand, has a better chance to emerge if we stick to physically definable terms, and one such term is "radiant energy." Let's see if radiant energy can help bring sensible answers to lingering questions.

Cooling, Heating, and Radiant Energy

When your hand encircles a container of water, it absorbs the radiant energy that the water emits (through the container). If the water emits a lot of IR, you interpret this as warmth; if there's not much IR generated, then you sense coolness. Your hand senses the radiation emitted, relays it to your brain, and presto! — you know if it's hot (**Fig. 10.7**).

Fig. 10.7 *Your hand senses the amount of radiant energy coming from the water.*

Thermometers sense in much the same way. They may employ either of two sensing modes: radiative or conductive. Those modes are presumed to differ, but I believe they are more similar than different. Let me explain.

Thermometers employing the radiative mode register the amount of radiant infrared energy received. They function much like your hand.

Conductive thermometers, by contrast, are said to "conduct" heat. The hard steel bulb "conducts" heat from outside, through contact with whatever lies outside the bulb. If that is water, then water's "thermal vibrations" transfer directly into the steel, which induces vibration of the steel's electrons, causing more distant steel electrons to vibrate, *etc*. Eventually, those vibrations propagate through to the mercury, whose consequent expansion lets you read the temperature from a scale.

This mode of sensing seems little different from the radiative mode. In both instances, radiant energy propagates through a medium and the amount is sensed and scaled to temperature. Essentially, your hand does the same: the tighter it grasps, the more intense the sensation (i.e., the water will seem "hotter").

Does Radiant Energy Carry Information?

Water emits radiant energy. Most of that energy comes from bulk water, but the EZ emits some energy as well. The wavelengths emitted from the EZ depend on its structure.

While the EZ structure has a generic aspect (Chapter 4), variants are anticipated. EZs build from surfaces with unique charge distributions. Those unique distributions will necessarily create variants of the generic EZ structure. Hence, the energy radiated from the EZ could contain surface-specific information.

If so, then EZ water may radiate information in the same way that TV station antennas radiate information. The radiated energy may be more than generic.

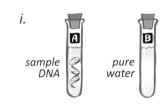

i.

sample DNA

pure water

What happens when water absorbs the radiated energy? If the radiated energy contains information, then we might expect that information to become blurred or lost. However, if some of the energy's vibratory modes induced new EZ structural variants, then some information could be retained. Any such retention would amount to nothing less than electromagnetically communicated structural information — a kind of water-based email.

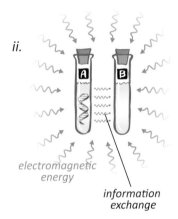

ii.

electromagnetic energy

information exchange

While any such communication might seem farfetched at best, stunning reports from Nobelist Luc Montagnier have lent credence to this kind of information transmission (see figure). Montagnier claims to have successfully transmitted DNA-structural signals to water. He first created an aqueous suspension of sample DNA. Then he placed the suspension in a sealed flask next to a second, similarly sealed flask of water. The flasks stood next to one another for an extended period, while he exposed both to a source of ordinary electromagnetic energy.

The newly "informed" water in the second flask was then combined with the raw materials required for DNA synthesis. This procedure created new DNA. The sequence of that DNA was not random: it was the same as the sample DNA in the first flask. Even though the two flasks had been well sealed and never came into physical contact, the information evidently passed from one flask to the other.[3,4]

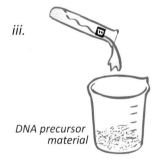

iii.

DNA precursor material

Initial responses to Montagnier's report have been skeptical. However, some scientists, persuaded by reports of electromagnetic transmission phenomena dating back to Gurwitsch[5] almost a century ago and by the more recent work of Benveniste[6], are in hot pursuit. At this writing two labs claim that they could confirm Montagnier's finding. It will be interesting to see what develops from these studies.

iv.

new DNA matching A

All of these sensing methods detect radiant infrared energy, and all of that energy arises from charge vibrations. I hope this provides an understandable linkage between radiant energy and the more vaguely defined concepts of heat and temperature.

We now press on, exploiting the concept of radiant energy. Under the umbrella of radiant energy lie two useful expedients that help in understanding water's properties:

• Protons released into bulk water during EZ formation constitute moving charges, which generate ample radiant energy. That radiant energy creates the sensation of warmth.

• EZ water generates relatively little radiant IR because of limited charge movement. The low infrared emission creates the sensation of coolness.

Armed with these expedients, we tackle some issues that have befuddled scientists and confused the dickens out of the rest of us. Here, we will consider two of them: (i) mixing and (ii) vortexing.

(i) Mixing

The Case of the Suspicious Heat and Missing Volume

A perplexing issue arises when substances mix with water. Beyond dissolution, you might expect that nothing particularly interesting will happen. However, ordinary mixing can bring serious consequences (**Fig. 10.8a**). Adding just a few drops of water to sulfuric acid can induce boiling, spitting, and sometimes even an explosion.

Fig. 10.8 *Mixing substances with water can bring unexpected consequences.*

That's not the only counterintuitive result of mixing. When liquids mix with water, the final volume doesn't always equal the sum of the two initial volumes (**Fig. 10.8b**): the volume can be higher, but more commonly it is lower. In extreme cases, the volume shortfall can be as much as 20 percent. Mixing solids with water can produce similar results: dropping a small number of sodium hydroxide pellets into a flask of water will reduce the original volume; to get back to that volume requires adding many more such pellets.

You can witness the volume phenomenon for yourself by filling a glass with water, just to the brim; then add salt. The water will not overflow even if you add salt to the point that a pile accumulates at the bottom of the glass. Volume seems to disappear.

Chemists know these phenomena well. According to prevailing theory, the burst of heat caused by adding a bit of water to sulfuric acid arises out of thermal contributions from each of the subprocesses underlying solvation; their sum yields the "heat of hydration." That heat makes the water hotter. The volume change phenomenon has a different explanation that depends on whether the mixing of molecules produces a better or worse intermolecular fit than prior to mixing.

While these explanations seem straightforward enough, there is no easy way to verify whether they are correct or incorrect. In the first phenomenon, the heat contributions are commonly surmised from the experiments themselves, rather than from independent observations. In the second phenomenon, determining how molecules fit like jigsaw puzzle pieces is not an exact science, to say the least. Hence, the correctness of these explanations remains uncertain.

When I began looking into these two phenomena, I was struck by an unexpected correlation: *heat emission and volume change seemed linked*. Whenever heat got emitted, shrinkage also seemed to occur; conversely, in cases where heat was absorbed, there'd be expansion. I wondered whether thermal and volume changes might therefore have a common origin.

The factor that had not previously been considered is the familiar one: the EZ. Mixing a solute with water seemed likely to change EZ content in one way or another: for example, if the substance were initially dehydrated, then adding water should promote EZ buildup, since EZ buildup is the very basis for dissolution (Chapter 8).

Suppose an increase of EZ occurred from the mixing of some substance with water. How might this affect "temperature" and volume?

• As EZs build, protons are released. Those proton charges move about, generating radiant energy; hence, the mixture should "heat up."

• Meanwhile, the mixture should shrink, because EZ density exceeds bulk water density (see Chapters 3 and 4). The shift from bulk to EZ water will therefore diminish the volume.

EZ growth, then, can explain heating and shrinking, at least theoretically. We set up experiments to test whether EZs really grew in those circumstances.

Resolving the Volume/Heat Enigma

We first sought reassurance. Heating and shrinkage seemed tightly linked, but we had to be certain. Therefore, we examined seven heat-generating dissolutions. In all of those cases, we confirmed the heat-volume linkage. (One case proved technically challenging, but other studies had already confirmed the linkage). Typically, the shrinkage took place within tens of seconds, and so did maximum heat evolution. The two phenomena tracked nicely, as we had anticipated.

Next, we tested whether those dissolutions generated EZs. Our standard test for EZ water is the presence of a 270-nm signature absorption peak. In all cases, the mixture showed this peak, or something close. Sometimes the peak could be shifted by 10 to 25 nm in one or the other direction; sometimes two sub-peaks could show up instead of the single peak; and some peaks could be weaker or stronger than others. But, in all cases, we found something close to the expected peak. A representative example is shown in **Figure 10.9**.

These results affirmed a linkage between heating and shrinkage, and they affirmed an EZ buildup whenever mixing created that heating and shrinkage. It appeared that we were on track — at least for those mixtures producing heating and shrinkage.

The flip side of the mixing coin is the less common outcome, cooling and expansion. To pursue that variant, we focused on a well-recognized example: the mixing of ammonium persulfate with water. Ammonium persulfate comes in powdered crystals. When the powder mixes with water in approximately equal parts, the volume increases beyond the sum of the two. We confirmed the expected expansion, although

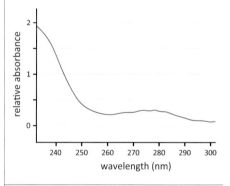

Fig. 10.9 *UV-Vis spectrometer recording of 50:50 mixture of HCl and water. A peak appears at 278 nm.*

technical difficulties prevented quantification of the effect. The temperature, on the other hand, was easy to track with a thermometer; it dropped by 8 °C. Hence, mixing behaved as anticipated: the mixture cooled, and it expanded.

A critical question was whether mixing diminished EZ content. **Figure 10.10** shows that it did: a broad peak of impressive magnitude centered approximately at 270 nm showed up initially. That peak became progressively narrower with each subsequent dilution.

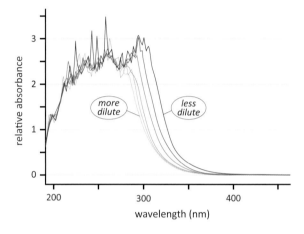

Fig. 10.10 *Ammonium persulfate dissolved in water. With increasing dilution (right to left), the area beneath the absorption peak diminishes.*

To understand the reason for these observations, it helps to know how powdered crystalline materials like ammonium persulfate are created. Although various methods exist, crystals commonly form following exposure to intense radiant energy (heat). Radiant energy builds EZs. We can presume, therefore, that the massive IR used to produce these materials builds substantial EZs and associated protons around each molecule. Those abundant likes and unlikes order the molecules, which then form crystals as the solution dries (Chapter 8, *box*). The seemingly dry crystalline powder contains no liquid water, but it contains EZ material in abundance, explaining its unusually robust 270 nm peak (**Fig. 10.10**).

Dropping such crystals into water reduces the crystalline order. The water provides a large reservoir into which the unlike charges can disperse. Proton dispersal diminishes the interactions between protons, and with diminished movement comes diminished infrared radiant output. Hence the solution feels cooler. Meanwhile, as the water invades, EZs diminish to smaller size, for their initially larger size depended on the intense radiant energy, which is no longer present. EZ diminution explains the narrowing of the 270-nm peak (**Fig. 10.10**).

Thus, thermal and volume features are explained: temperature drops because radiant output diminishes; and volume expands because dense EZ water converts to less-dense bulk water.

The EZ paradigm seems to suffice not only for explaining the heating/shrinkage regime but also for explaining the cooling/expansion regime. To quantify and systematize these observations, more comprehensive studies are needed. Perhaps it will then become clear why the innocent act of dribbling water into a vat of sulfuric acid can produce an explosion.

Vodka and Viscosity

Dmitry Mendeleev, the Russian scientist who brought us the periodic table, also studied the mixing of ethanol and water. Like most mixtures, this one produces heat and shrinkage. Mendeleev noted another attribute: an increase of viscosity, by as much as three times. The reason has remained obscure, but the EZ's high viscosity (see **Fig. 3.17**) might explain the result. Ethanol forms EZs much like water;[7] therefore, a mix of ethanol and water might produce intermingling EZs exhibiting substantially elevated viscosity.

While that's speculation, what's beyond speculation is the practical result: because he found the highest viscosity at a 40:60 mixture of ethanol to water, Mendeleev considered that particular ratio ideal for making vodka. That's the ratio used to this day. The high viscosity lends the drink a satisfying "body," enough perhaps to explain why Russians consume so much of it.

Dmitry Mendeleev
1834–1907

Concrete

Following the standard procedure, you add water to concrete powder, mix vigorously, pour, and let the poured mixture set for a day or two. The concrete eventually hardens. As it hardens, you notice that it emits heat. Why so?

Adding water to the mix creates workable putty, much like wet sand. The putty-like consistency arises in the usual way: EZs build around the wetted particles;

released protons then mediate the like-likes-like attraction. So particles begin to cling. Particles first cling weakly, with putty-like consistency, and then progressively more firmly as charges continue to build and attractive forces grow higher.

Heat is an expected accompaniment. EZ buildup yields a progressive release of protons. As those concentrated protons move about, they generate abundant radiant energy — which is sensed as heat.

Meanwhile, that radiant energy helps bring the process to completion. That energy promotes further EZ growth, more released protons, and hence increasingly strong attractions. That's what makes the concrete hard throughout. This facilitating principle may apply not only to concrete but also more generally.

(ii) Vortexing

Why Vortexing Induces Cooling

We next consider the second paradoxical phenomenon: vortexing. Why should vortexed water feel cooler than non-vortexed water?

Vortexed water became a subject of discussion following the work of the legendary Austrian naturalist, Viktor Schauberger. Schauberger spent much of his life studying water. Lifelong observations convinced him that vortices lent water a special "vitality." Schauberger considered water taken from fast-running, vortex-filled streams more "alive" than stagnant water, which he considered dead. He also considered vortexed water as cooler.

Schauberger's work, a century ago, followed in the footsteps of another legendary Austrian, Rudolph Steiner, whose diverse spheres of endeavor included agriculture. Steiner invented the so-called biodynamic farming technique, whose central feature is water vortexing. To this day, some farmers and fruit growers tout this vortexed water as producing crop yields of stunning abundance without fertilizers.

Vortexing remains surprisingly popular among some groups today. Yet, beyond the empirical observations demonstrating its variously touted features, scant fundamental evidence can be adduced to lend a scientific underpinning.

I would speculate that at least some of the features of vortexing derive from the creation of EZ water. Preliminary experiments have confirmed an absorption peak at 270 nm (**Fig. 10.11**), implying that EZs are indeed present. Additional experiments are underway.

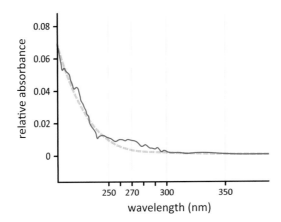

Fig. 10.11 *Spectral measurements of vortexed water. Preliminary observations show EZ growth.*

If EZ water is present, as preliminary observations imply, then what Schauberger referred to as "vitality" may correspond to energy, for exclusion zones bear potential energy. As for the cooling: if vortexing transforms some bulk water to EZ water, then the ensemble should feel cooler because EZ water radiates less energy than bulk water. Less radiant energy translates to coolness.

(The reader may notice a possible hitch: the freshly created EZ generates protons, which could counter the cooling effect by generating radiant energy; however, any such protons will quickly dilute because of the large water volume lying beneath the vortex; this would minimize those protons' contribution to heat. The net result of vortexing ought to be cooling.)

Why vortexing should induce EZ buildup may seem less mysterious if you recall that EZ water contains more oxygen than bulk water. Swirling puts water into continuous contact with oxygen, both from the air above and from bubbles trapped beneath. Thus, water continuously mixes with oxygen, enabling EZ buildup. Furthermore, substances moving past air inevitably become negatively charged (see Chapter 9), and negative charge also promotes EZ buildup (**Fig. 5.8**). All of this seems to suggest that investigating how vortexing might build EZs and, therefore, how it induces cooling, constitute worthwhile goals.

Resolving both the vortex-cooling and dilution paradoxes helps to highlight the chapter's key theme: understanding water's thermal features requires returning to fundamentals. At the very core of those fundamentals is charge. More charge movement generates more radiant energy, which makes the emitting substances seem "hotter." Less charge movement generates less radiant energy, making them feel "cooler."

Adhering to this fundamental concept has just helped resolve several anomalies and will serve us well as we launch into an exploration of other natural phenomena.

Summary

Water's thermal features have remained riddled with anomalies and paradoxes. To resolve those paradoxes, we stepped back from standard explanations based on heat and temperature and retreated to more fundamental approaches. In particular, we focused on radiant energy.

Radiant energy originates from charge movements. Back-and-forth displacements of charge give rise to electromagnetic waves, which propagate through materials and often emerge with a different character. In the case of water, the most relevant wavelengths lie in the infrared range. Water both radiates and absorbs significant amounts of infrared energy because of the water molecule's atomic structure. Thus, infrared wavelengths take on a particular relevance, which helped us to understand how heat and temperature relate to radiant energy.

A second feature of our approach was to recognize the contribution of exclusion zones in producing radiant energy. EZ buildup generates protons, whose movements generate plenty of IR. We sense this infrared energy as heat. Once the EZ stops building and stops contributing protons to the bulk water, what counts most is the fraction of the water that has become EZ water. Larger fractions of EZ water generate relatively less IR output, which may be sensed as cooling.

These features helped explain why mixing substances with water can lead to heat transients and volume changes. They also helped resolve the chapter's opening paradox: why vortexed water feels cool.

All of these findings arose by considering radiant energy, and by avoiding reliance on the terms heat and temperature. While those latter terms may be indispensable for everyday use, their ambiguity makes them unreliable concepts for advancing scientific understanding.

To attain that understanding, we continue our exploration using the vehicle of radiant energy. We will see how the notion of radiant energy can help us to understand water's common everyday behaviors.

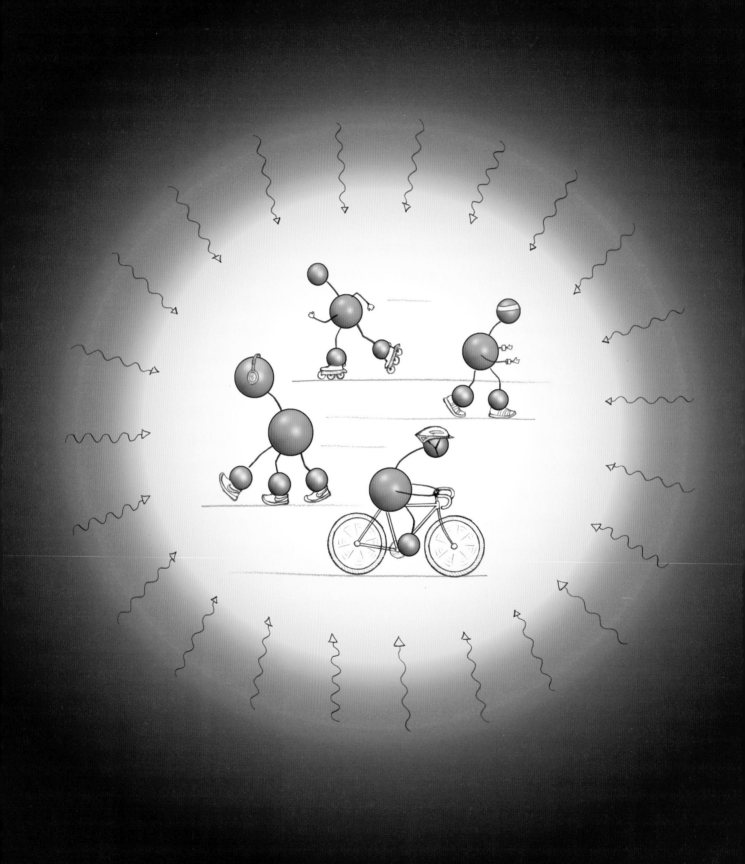

11 Osmosis and Diffusion: They Don't Just Happen

A famous Garfield cartoon shows the overweight cat with a stack of books piled high upon his head. Garfield claims, "I'm learning by osmosis." Osmosis offers the lazy a modicum of specious hope for painless knowledge transfer: for information to seep from the reservoir of knowledge to one's waiting brain.

Real osmosis, the prototype for this metaphor, is the process that transfers water from a place where it is concentrated to a place where it is less concentrated. The water moves. Osmotic water movement played a central role in Einstein's understanding of Brownian motion, and this chapter fulfills my earlier promise (Chapter 9) to reconsider the osmotic mechanism.

Dealing with osmosis without dealing with diffusion seems restrictively narrow, for these phenomena mirror one other. Diffusion involves particle or molecule movement through a fluid, while osmosis involves fluid movement toward particles or molecules (usually through a membrane). Loosely speaking, these phenomena are opposites. Both of them reduce concentration gradients by moving substances to regions where they are less concentrated. They are nature's principal vehicles for moving things around.

This chapter addresses questions about how these processes work. Do they happen spontaneously as a consequence of some fundamental natural law? Or does some underlying energy drive them, as wind drives a windmill?

Diffusion: Osmosis Inside Out

You throw a pinch of salt into the chicken soup (with or without matzoh balls). The salt spreads. Technically, the salt diffuses. The soup eventually becomes uniformly tasty.

Diffusion theory follows the prevailing theory of Brownian motion. By dint of "thermal" motions, each molecule will bounce around in a so-called random walk, ensuring that the molecules will spread out. We could liken those molecules to the drunken sailors emerging from a saloon: if confined within a fenced zone (and barring distractions) the sailors will eventually distribute themselves in a statistically uniform manner (**Fig. 11.1**). The salt will likewise spread throughout the soup.

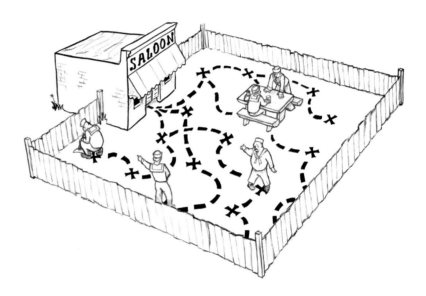

Fig. 11.1 *Random walks will eventually lead to a statistically uniform distribution spread out over a defined space.*

While random walks may take place as described, those walks require energy: if diffusion is the collective result of Brownian motions, and Brownian motions need energy (Chapter 9), then so must diffusion. It cannot be otherwise. The diffusion process may seem passive, but some kind of energy must drive it.

Standard diffusion theory does not include an external source of energy. It describes the diffusive spread in terms of the diffusion constant, **D**, which depends on various physical factors (see *box*, Chapter 9), but not on input energy. It theorizes that diffusive flow occurs spontaneously.

As often as not, however, the theorized spread fails to agree with the observed spread. A classic example comes from the diffusion of charged polymers in aqueous solution: diffusion occurs in both ordinary and extraordinary modes.[1] The mode that prevails depends on salt concentration. A tiny reduction of salt concentration can cause an abrupt shift from ordinary rapid diffusion to extraordinarily slow diffusion. The reason for this bistable behavior has remained elusive for those steeped in the framework of classic theory.

Discrepancies between observation and theory often lead to the introduction of such expedients as "sub-diffusion" or "super-diffusion." Proteins, for example, are said to exhibit sub-diffusion,[2] whereas particles in meteor trails are said to exhibit super-diffusion.[3] These terms merely emphasize that standard diffusion theory does not work as consistently as one might hope. The conventional formulation misses something.

Even common everyday phenomena illustrate this limitation. An example is the mixing of river water with ocean water. Rivers empty their water into the sea, and standard diffusion theory predicts that the two bodies of water should mix readily. However, the theoretical prediction does not bear out: in some places, salt water and fresh water can remain separated practically indefinitely.[w1] Even different salt waters don't mix easily: near the resort town of Skagen, Denmark, where the Baltic Sea meets the North Sea, the merger line remains permanently visible.[w2] An example of this merger line is shown at the right.

Merger of Baltic and North Seas create permanent boundary line.

Curious about these well-recognized deviations from theory, we followed up with experiments of our own. We poured a saturated salt solution into a beaker and then topped up the beaker with pure water. The pure water on the top contained dyes or microspheres, allowing us to see how the top and bottom solutions mixed. No obvious mixing occurred for many hours; and sometimes days passed before the two became appreciably mixed. The same thing happened in the upside-down experiment, where we poured the salt water over the pure water: still no obvious mixing. The upside-down result implies that density differences cannot explain the sustained segregation. Diffusion should have moved the molecules and particles around, but they didn't easily mix, just like river and ocean waters don't easily mix.

We followed up by injecting a blob of dye into a corner of a chamber, to see how it spreads. Some of the results shocked us. Injected into a chamber of pure water, the dye diffused more or less as the diffusion

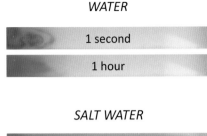

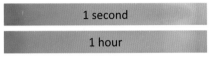

Fig. 11.2 *Diffusion of methylene blue dye. Chamber length 7.5 cm. In pure water, the dye diffuses as anticipated; in salt water (4 M KCl), the dye diffuses extremely rapidly.*

equation predicts, rather slowly (**Fig. 11.2**, *top*). But when the same dye was injected into a concentrated salt solution, it diffused so rapidly over a thin surface layer that you couldn't follow it by eye; within one second it covered practically the entire surface (*bottom*). Once that happened, the dye hardly diffused downward into the body of the solution, even after a week.

Stunned by these results, we went on to check whether the differences might have arisen from some quirk of the particular dye or salt. Several dyes gave similar results, as did the substitution of algal cells for the dye. Nor did it matter whether the salt was potassium chloride or a variety of other common salts. We still saw a qualitative difference between diffusion in pure water and diffusion in salt water.

These experiments showed how grossly real diffusion and theoretical diffusion can differ. The diffusion equation may be simple, easy to apply, and adequately predictive under a restricted set of circumstances. However, that equation falls short of providing a general explanation for the spread of molecules.

What went wrong with the theory?

The diffusion equation came from the concept of thermal motion: substances diffuse as a result of jittering about. However, if external energy drives the jitter (Chapter 9), then external energy must also drive diffusion. So we should not expect the diffusion equation to work properly without the inclusion of an energetic term. Including that energy in a modified theory could constitute a useful first step (**Fig. 11.3**, *top*), although it might not be trivial.

A second step involves accounting for forces that "distract." Localized charge will distract the solutes from their normal behavior in the same way that an alluring woman will distract most drunken sailors (**Fig. 11.3**, *bottom*). The solutes will move in predictable directions, either approaching the localized charge or fleeing from it.

Thus, diffusion is not simply a matter of temperature, particle size, and the medium's viscosity, as the classical diffusion equation (*box*, Chapter 9) theorizes. *To predict properly, the formulation must take account of the absorbed incident energy, along with any distracting charges that may be present.* Only then will the formulation begin to reflect reality.

We can envision how to approach the development of a more adequate theory. It must start by focusing on the central role of external energy. Absorbed energy builds EZs around particles and molecules. Those EZs separate charges, which then become subject to the whims of the charged distractors. Positively charged distractors will draw negative EZs and any free OH^- groups; negatively charged distractors will draw positive EZs and hydronium ions. Higher incident energy will intensify those draws. The bottom line is that these energy-based attractions are more than secondary phenomena: *energy-driven, charge-based forces largely govern diffusive movements.*

Solute diffusion is then much like drunken sailor diffusion: if you want to determine the sailors' instantaneous locations you need to take account of their energy and their distractions.

The failure of pure water to mix easily with salt water may arise as a consequence of the like-likes-like mechanism. Salt molecules envelop themselves with EZs.[4] As those EZs build, they generate opposite charges beyond — creating like-likes-like attractions. With high enough concentrations of salt, those EZs pack into ordered arrays, much like colloid crystals (see Chapter 8). Indeed, a long line of optical scattering evidence confirms that dissolved salt molecules pack into massive water-containing clusters.[5] Those clusters may resemble colloid crystals.

At high salt concentrations, then, EZ water should dominate the lattice space. EZs exclude most everything — including even bulk water (see **Fig 11.6**, *below*). Thus, any water positioned next to the salt lattice should remain separated, even over the long term. Such persisting separation would explain the difficulty in mixing river water with salt water.

water

salt water

Water cannot easily penetrate into a "crystal" of salt water.

Osmosis: Another Uncertain Phenomenon

Having considered diffusion, we now flip the coin and consider osmosis — the diffusive movement of water. Since water is a molecule, it should behave like other molecules: its diffusive movements should conform to the same principles, including those derived from external sources of energy and charge-based distractions.

Conventional osmosis theory does not consider energy, but it does at least consider distractors. The distractors are the solids. Suspended or dissolved solids are said to "attract" the water molecules, which then diffuse toward the solutes. The attraction is not conceived as charge based; it is conceived in terms of "concentration": the water molecules move from regions of higher concentration (such as pure water), to regions of lower concentration where the water intersperses with solids. Thus, water draws toward the solutes.

Following this formulation, scientists including Einstein have adopted what might be called a "Garfieldian" approach to osmosis: a passive flow of water, much like the passive flow of knowledge. The water concentration evens out, just like the particle concentration evens out by diffusion. However, *osmosis cannot be passive*. If diffusive movements underlie osmosis and absorbed energy drives diffusive movements, then absorbed energy must play some role in osmosis. Absorbed energy's contribution cannot simply vanish for theoretical convenience.

So why then does the water move?

Most of us were taught along the following general lines: osmosis occurs because all things tend towards equilibrium. The water "tries" to equalize its concentration. Thus, when an intervening membrane separates two compartments, molecules will move toward the compartment with more solutes (**Fig. 11.4**). This supposedly happens because water molecules bounce around in endless thermal motion; if they can pass through the membrane, then the concentrations in the respective compartments will try to even out, raising the water level in the solute-containing compartment.

This account is hardly the only one. Scientists have debated the mechanism of osmosis for over three centuries, offering an array of proposals that still lie on the table. Those proposals include what might be called a water concentration theory, a solute bombardment theory, a solute attraction theory, and a water tension theory — as well as a variant of the latter to be advanced in a forthcoming book by an Australian colleague, John Watterson. Meritorious arguments support each proposal, but no proposal seems to explain everything; hence, none has gained universal acceptance. The bottom line: we still have no clear explanation for the mechanism of osmosis.

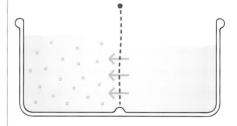

Fig. 11.4 *Standard experimental setup for observing osmosis. The membrane can pass water, but not solutes. The water moves from the low solute concentration compartment to the high solute concentration compartment, raising the water level on the left.*

Osmosis as an Epiphenomenon

One possible reason why the osmotic mechanism has remained elusive is that nobody could envision a role for absorbed energy in the process; energy-driven movement was a foreign concept. Another reason is that, in the standard membrane experiment used for studying osmosis, nobody could imagine that the separating membrane itself would turn out to be a critical feature. The membrane had seemed nothing more than a passive barrier, blocking solutes but allowing the smaller water molecules to pass from one chamber to the other.

We now know that hydrophilic membranes bear exclusion zones. If EZs form on both sides of the membrane, then both compartments will be filled with protonated water. And, if the concentrations of protonated water differ in the two compartments, then a proton gradient would straddle the membrane. As I will show, any such proton gradient would drive hydronium ions through the membrane, shifting water from one compartment to the other.

So the membrane could be a central protagonist in the osmotic drama. It could create a hydronium ion gradient that drives osmotic water flow. On the other hand, this hypothetical mechanism raises a thorny question: any EZ lining would constitute a barrier through which water molecules would need to pass if they attempted to flow from one compartment to the other. How can water pass through the EZ liner?

We sought to test the proton-gradient hypothesis, hoping to resolve that thorny penetration issue in the process. To test for proton gradients, we used the standard apparatus comprising two contiguous chambers separated by a standard osmosis membrane, as depicted in **Figure 11.4**. We mounted the commonly used membrane, whose pores are presumably large enough to pass water but not solutes. With this apparatus in place, we could track water movement toward the side with solutes. We could also examine both sides of the membrane using a microscope.

The expectations detailed above were confirmed.[6] First, we confirmed the presence of exclusion zones straddling the membrane. We saw them not only next to the cellulose-acetate membrane used routinely for osmosis experiments, but also next to a Nafion membrane, which could produce similar osmotic flows.

In both cases, however, the two EZs differed in size (**Fig. 11.5**). The right-hand surface of the membrane, exposed to pure water, should bear an exclusion zone of standard size, which it did. The left surface faced a salt solution. Salt diminishes EZ size.[7] Therefore, the EZ on the left side of the membrane was much smaller.

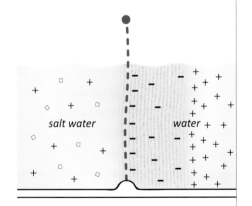

Fig. 11.5 *Standard osmosis experiment, with exclusion zones and protons distributed asymmetrically around the dividing membrane.*

190

What about protons? The right-hand chamber should contain numerous protonated water molecules because of the large EZ abutting the membrane. The left EZ, being much smaller, should generate few protons. Voltage measurements across the membrane confirmed that the right chamber was substantially more positive than the left. So, the anticipated electrical gradient proved real.

Now, envision the fate of those right-side hydronium ions. As positively charged water molecules, hydronium ions repel one another and want to escape from the right chamber. The obvious destination is the left chamber, which has fewer positive charges and therefore sits at a lower electrical potential. The protonated water molecules should therefore flow leftward, creating the "flow of osmosis" which should continue until the driving gradient vanishes.

The scenario above describes a role of the membrane in the osmotic draw. It shows how the asymmetry drives the flow of protonated water. However, the membrane is not the sole drawing agent; the salt molecules themselves facilitate the draw, and I will return in a moment to explain how.

Meanwhile, this model has profound energetic implications. Absorbed radiant energy fuels EZ buildup, which separates charges, which propel the flow. Hence, osmosis is *energy driven*. This conclusion contrasts sharply with the understanding implicit in all of the competing theories, that osmosis is a passive process requiring no energy at all. Osmosis *does* require energy. Getting something for nothing is an expedient that not even nature can accomplish.

Dams with Holes

Regarding the osmotic flow, we must still face the problematic issue of how those hydronium ions work their way through the seemingly impenetrable EZ barrier.

Direct penetration is unlikely because of the EZ's narrow pore size. The unitary hexagon is barely large enough to pass a water molecule; the actual mesh opening is even more diminutive because contiguous EZ planes lie out of register (**Fig. 4.15**) — effectively narrowing the mesh size by a factor of about three. A water molecule should not pass through.

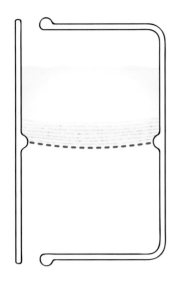

Fig. 11.6 *Water fails to penetrate the EZ.*

An accident confirmed that expectation. An undergraduate student carrying out an osmosis experiment using an apparatus similar to that of **Figure 11.5** returned to the lab one morning to find that he had forgotten to clean up the night before. The salt water had leaked out of the chamber because of a poor seal; yet the chamber on the other side of the membrane remained full. The intervening membrane had not permitted the pure water to penetrate. This surprised us at first, because water evidently does pass through the membrane during the actual osmosis experiment.

Follow-up experiments confirmed the same result: With the salt side empty from the start, and the other side filled with pure water, we still saw no transmembrane flow. Even when we tilted the apparatus 90° to get an assist from gravity, the water failed to pass from top to bottom (**Fig. 11.6**).

The paradox remained: although the EZ mesh is apparently too fine to allow water to pass, water demonstrably flows from one chamber to the other during osmosis. We were completely befuddled.

Visual observation finally resolved the paradox (**Fig. 11.7**). While EZs ordinarily cover the full area of the membrane, in the osmosis experiment they do not: microscopic exploration uncovered surprisingly large breaches.[6] Those breaches amount to portals through which the water can easily pass — just as water can pass through holes in dikes. Paradox resolved.

Fig. 11.7 *EZ breaches observed microscopically* (left). *To create the breaches, positive charges invade the negatively charged EZ, eroding it locally* (right).

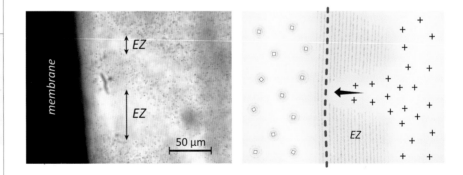

What creates those breaches? Locally developing EZs generate protons. Those protons flow back through the membranous regions in which EZs have not yet begun to build, toward the left chamber (**Fig. 11.7**, *right*). This flow inhibits EZ buildup in those latter areas; hence, EZ buildup remains patchy, as observed.

EZ breaches are not unique to osmosis. We see them also in the EZs adjacent to certain metals (see Chapter 12); presumably they exist around cells, where molecules need to pass through an otherwise enveloping EZ in order to leave or enter. Indeed, charge gradients spanning from inside to outside the cell could drive flow in the same way as they drive flow in osmosis.

The Salt Draw

Another issue is the role of the salt molecules. EZs build around salt (or other solute) molecules. We might ask what role those EZs play in the osmotic draw.

Consider **Figure 11.8**. With positive hydronium ions abutting the membrane on the right side, salt particles should cluster on the left side, because they bear negatively charged EZs. It is a matter of simple attraction. Those negative EZs remain stuck in place; they cannot pass through the membrane. Their presence draws the hydronium ions leftward. This flow maintains the breaches in the right-side EZ, ensuring continuous hydronium flow through those breaches.

Hence, the hydronium ions move from right to left not only because repulsion pushes them out of the right-hand chamber, but also — and perhaps mainly — because negative EZs exert a pulling force. That pull ensures their leftward flow.

This latter mechanism leads us to a general statement about osmosis. In the narrative above, we dealt with the standard experimental configuration for studying osmosis: a membrane separating chambers of low and high salt concentrations. That arrangement created a lot of hydronium ions on one side of the membrane and lots of negative charge on the other side. It caused hydronium ion flow, which raised the water level on the saltier side (**Fig. 11.4**).

However, that particular configuration is only one of many in which charges are separated. Any separated pool of hydronium ions will bring flow: those positively charged water molecules will inevitably flow toward negatively charged EZs. That flow is the flow of osmosis.

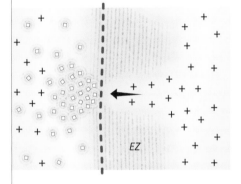

Fig. 11.8 *EZ clustering against membrane draws hydronium ions leftward.*

Thus, membranes are unnecessary — the membrane is merely a convenient artifice for separating positive hydronium ions and negative charge. Membrane or no membrane, *the central protagonist of osmosis is the hydronium ion, created by the absorption of external energy.* Hydronium ions always move toward negative charge.

A question arises as to why these electrical charge gradients had never before been detected. With three centuries of osmosis study, you'd think that someone might have observed them. In fact, someone did: In now-classical experiments carried out a century ago,[8] Jacques Loeb established the presence of an electrical potential difference from one side of the osmotic membrane to the other.

Sadly, that pivotal finding seems to have been forgotten amidst the rush of molecular-scale experiments. The electrical potential difference is clearly present; as Loeb aptly recognized, its presence must surely be a factor in driving the osmotic flow.

So osmosis occurs in much the same way as diffusion (**Fig. 11.9**). Charge gradients drive both types of flow — gradients built by absorbed radiant energy.

Fig. 11.9 *Summary of diffusion and osmosis mechanisms, illustrating common features.*

DIFFUSION

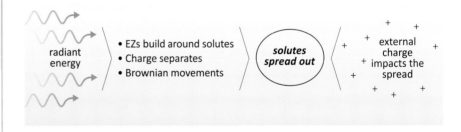

OSMOSIS

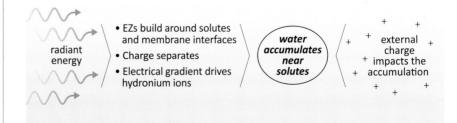

Diapers and Gels

Osmosis operates in everyday life. Common examples that illustrate the principles just elucidated can be found in gels and diapers.

Gels hold enormous amounts of water. Gelatin desserts comprise 95 percent water, while some laboratory-based gels may comprise up to 99.95 percent water.[9] Diapers (nappies) have similar features. Fortunately for our convenience, diapers can absorb many times their weight of water. This holding capacity makes sense because EZ water can cling to the diapers' internal hydrophilic surfaces, and those EZs can be of substantial size.

The invasion begins when the gel's (or diaper's) dry hydrophilic meshwork gets exposed to water. The water does more than just fill the open mesh spaces; it expands the meshwork. Within seconds or minutes the meshwork swells, sometimes to impressive dimensions. Osmosis is evidently at play, for water flows toward solids.

How does this massive flow happen?

When the dry gel is plunged into water, superficial meshwork layers and protruding polymer strands immediately begin hydrating. EZs build. Hydronium ions begin accumulating beyond. If the matrix bears negative charge, then those hydronium ions will begin their inward rush, into the matrix proper.

Negative charge is typical of hydrophilic matrices. There are several reasons: one, because the polymers themselves commonly bear negative charge; another, because even "dry" polymers contain some tightly held EZ water that is practically impossible to remove. Paper (cellulose) for example typically contains 7–8% water; even several days in a dry oven cannot remove that water. Thus, ample negative charge in the matrix draws those hydronium ions inward, inevitably dragging contiguous dipolar water molecules along for the ride. So the matrix begins filling with water and hydronium ions. This is osmotic flow: hydronium ions flowing toward negative charge.

You can envision the progression of events. The penetrating water provides the raw material for building more EZ layers; as those EZs build from incident energy, they release protons; the protonated water molecules penetrate more deeply into the negative matrix, dragging along more water, and so on. Soon the entire gel fills with water. By that point, all matrix surfaces bear substantial EZs, leaving the intervening pockets filled with protonated bulk water.

What finally stops the osmotic inflow? Certainly not matrix-charge neutralization, for even well-filled gels remain negatively charged. Microelectrodes stuck into those gels reveal negative electrical potential (e.g., **Fig. 4.7**). Thus, even swollen matrices should retain the capacity to draw protonated water inward.

By this argument, you'd think the gel could expand unendingly; however, mechanical constraints limit its growth. An elastic mesh can accommodate a good deal of water before the elastic limit is reached and flow stops. A stiff mesh will curtail the water accumulation sooner. The end point of filling is reached when the mechanical resistive force balances the osmotic drawing force.

At that stage, the gel contains a lot of EZ water. The pockets between those EZs contain protonated water. That protonated water sticks to the negative EZs and remains constrained within the gel unless pressured out. Your gelatin dessert should be of this ilk, and so should your child's wet diaper.

An amusing incident confirmed the nature of the osmotic draw. On a recent trip to Pisa, a colleague took me to a favored local restaurant. Once seated, we watched the waitress perform her act of flamboyance. She ceremoniously plunked a small white cylindrical blob onto each of our plates (**Fig. 11.10**, *top*). Since the blobs lay on our plates, I reflexively presumed they were edible — perhaps some exotic local seafood. Exotic they were: when the waitress dribbled some water on the blobs, they sprang to life as though the water were holy: they magically grew to more than five times their initial height (**Fig. 11.10**, *second panel*).

The blobs were not food after all. They turned out to be ribbons of fibrous meshwork rolled into tight, squat cylinders. When unraveled, the hydrated ribbon could be seen as nothing more than wet tissue,

Fig. 11.10 *Hydration-induced swelling. Hydrophilic meshwork is initially dry* (top). *Adding water expands the meshwork* (second panel). *The unrolled meshwork is now ready for hand wiping* (lower panels).

Injury and Swelling

Osmosis plays a role in cell function. Since the cell is packed with negatively charged proteins, the cytoplasm should generate an osmotic draw (commonly referred to as osmotic pressure) similar to the osmotic draw generated by gels, tissues, and diapers. Physiologists know that it does.

A peculiar feature of cells, however, is their relatively modest water content. Compared to 20:1 for many common gels, the water-to-solids ratio in the cell is only about 2:1. The cell's many negatively charged macromolecules should generate a strong osmotic draw; yet the water content in the cell remains low. That limited water content may come as a consequence of the macromolecular network's stiffness: cellular networks typically comprise cross-linked tubular or multistranded biopolymers. The resultant stiffness prevents the network from expanding to its full osmotic potential.

If those cross-links were to disrupt, however, then the full power of osmotic draw would take effect; the tissue could then expand enormously. In experiments with single muscle fibers, we could see this expansion all too frequently: when the experimenter's forceps inadvertently slipped, local damage created a knot that could rapidly expand up to some ten times the diameter of the rest of the fiber. The banding structure was locally disrupted, and water evidently invaded from outside to swell the tissue. At that stage, we knew that the experiment was ruined.

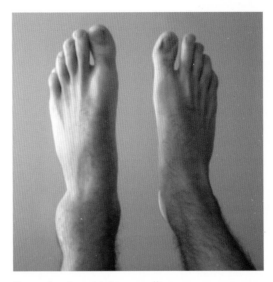

Example of post-injury swelling.

Swelling occurs similarly when body tissues are injured, especially with dislocations. The injury disrupts fibrous macromolecules and cross-links, eliminating the restraining forces that keep osmosis at bay; osmotic expansion can then proceed virtually unimpeded.

The reason why swelling can be so impressive is that the disruption occurs progressively. Breaking one cross-link results in higher stress on neighboring cross-links; so disruption progresses in a zipper-like fashion. When that happens, the osmotic rush of water into the tissue can continue practically without limit, resulting in the enormous immediate swelling that is often seen. The tissue will return to normal only when cross-links repair and the matrix returns to its normally restraining configuration.

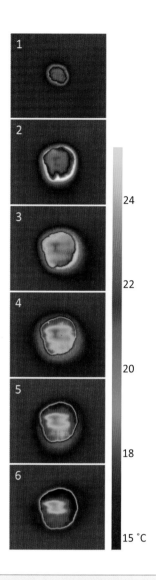

Fig. 11.11 *Sequence of infrared images of a water droplet spreading on a napkin. During the spread, the outer rim contains a "hot" region.*

designed for wiping your dirty hands (*bottom panels*). The diaper-like material had been cleverly packaged into a compact cylindrical form, poised for action.

The meshwork result held particular interest because we had recently begun examining hydrating meshworks using an infrared camera. We would place a droplet of water on a flat tissue mesh. As the water spread, the infrared camera would always detect a "hot" zone at the water's leading edge; i.e., the leading edge generated abundant infrared energy (**Fig. 11.11**). That observation finally made sense: moving charges generate infrared energy. The advancing hydronium ions — the moving charges at the leading edge of the invading water — bore responsibility for the high infrared emission. Those charges led the way.

So charges drive the penetration of water into hydrophilic polymer networks — just as they drive the penetration of water toward solutes. Both of these osmotic phenomena involve hydronium ions drawing toward negative charge. And both are fueled by external energy, which separates the charge responsible for the osmotic draw.

Summary

Salt sucks! This crude expression has become a familiar artifice for remembering what goes on during osmosis, where salt attracts water. We found, however, that the salt doesn't really attract the water. Salt (and other solutes/particles) build negative EZs around them, thereby attracting positive hydronium ions. That electrical gradient drives the osmotic flow.

Osmosis, then, is a process secondary to the absorption of incident radiant energy. That energy separates charge, which drives the flow. Osmosis is *not* the elementary force of nature that the classical Brownian motion formulation presumes (Chapter 9).

Like osmosis, diffusion is a mixing process. Diffusion deals with movement of solutes rather than movement of solvent. The driver is once again external energy, which separates charge. Those separated charges drive the excursions responsible for mixing.

Thus, the issues surrounding osmosis and diffusion are similar. Both require energy, and energy-driven charge separation is central to both types of movement. You might think of osmosis and diffusion as natural consequences of the sun's energy.

If energy and charge strike you as appearing frequently in these pages, then your perception is accurate. The next section continues in this vein. It begins with explorations of simple everyday phenomena ranging from why ice is slippery to why your joints don't squeak. Separated charges provide surprisingly simple explanations.

This section builds on basics. It considers the extent to which the previously developed concepts can help us develop a fuller understanding of everyday phenomena, ranging from the power of batteries to the merging of bubbles.

Many of these understandings should be well grounded in available evidence. A few may be tentative: so much darkness shrouds the terrain of understanding that fresh light cannot possibly illuminate everything. Where understandings remain speculative, I will post reader alerts. Please keep an eye out for the graphic below.

SECTION IV

Aqueous Forms in Nature

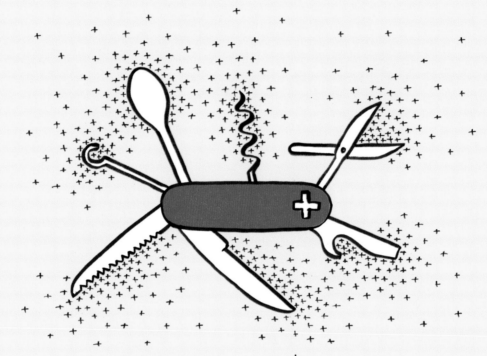

12 The Power of Protonated Water

Half asleep at the back of the chemistry lecture hall, I just didn't get it. I was an undergraduate. The subject was the proton.

Everyone knew that protons were positively charged particles tightly clustered at the center of the atom, but the professor began telling us they could also be donated. Acids, he said, donate protons. So protons must also be stuck somewhere on the periphery of the atom — which was all right, except that I couldn't see how that positive proton, already latched onto a peripheral electron, could be so easily "donated." You'd think opposite charges would stick like magnets.

Donating to the cause.

The professor went on to expound on all the wondrous things that those protons could do, from driving reactions to mediating battery function. But I still couldn't wrap my head around the logic. Lacking the confidence to risk divulging my stupidity, I kept my mouth shut, condemning myself to a state of lingering ignorance. A chemist I'd never be! Those protons seemed endowed with all kinds of mysterious functional powers, but to me, the nature of those powers seemed elusive.

Finally, after many years, the seeds of understanding have begun to sprout. I can see more clearly what gives protons their vaunted power. The magic reduces to three simple features: First, they are abundant; EZ buildup creates colossal numbers of protons. Second, those protons quickly latch onto water molecules, the molecular fusions creating charged water molecules brimming with potential energy. Third, those now-charged water molecules obey the laws of physics: they move toward negative charge and withdraw from positive charge.

So proton power boils down to electrostatics. Electrostatic forces play a key role in osmosis (Chapter 11). Here, we extend that simple electrostatic concept to explore whether a half dozen everyday phenomena that have resisted easy explanations can be understood in terms of electrostatic attractions and repulsions.

Fig. 12.1 *Friction analogy. When the upper "mountain range" slides past the lower one, the resistance creates friction.*

(a)

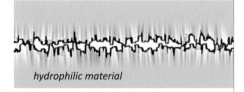

hydrophilic material

(b)

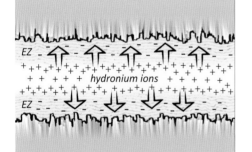

Fig. 12.2 *Water and friction.*
(a) *Hydrophilic surfaces may contain friction-generating asperities.*
(b) *With water lodged in between, the surfaces are pushed apart by repulsive forces between hydronium ions.*

(i) Proton Repulsion Reduces Friction

Rub one smooth surface upon another. The resistance you feel arises from "asperities" — microscopic protuberances projecting from the nominally smooth surfaces. The protuberances run into one another, creating friction (**Fig. 12.1**).

Take the familiar example of sandpaper. Rubbing two sheets of sandpaper together exemplifies the situation depicted in **Figure 12.1**. Sand collides with sand, and the frictional coefficient is said to be high. Rubbed on smooth wood, however, the sandpaper has a lower frictional coefficient; wet soap on your skin has an even lower coefficient.

From **Figure 12.1**, you can surmise an easy way to reduce friction: push the respective surfaces apart so the peaks don't collide as much. For example, suppose that the interfacing materials are made of hydrophilic polymers. Because strands of polymer may protrude from both surfaces like bristles of a brush, the coefficient of friction will be substantial; the strands cannot help but run into one another. However, if water is infused between those polymeric surfaces, then each surface will sport an exclusion zone, which will push the material surfaces apart. This distancing reduces friction.

The polymer-water story does not end there. Each negatively charged EZ will generate hydronium ions (**Fig. 12.2**). Those hydronium ions repel one another, creating a repulsive force that can push the respective surfaces apart. If measures are taken to keep those hydronium ions from escaping the gap zone, then the asperities could remain fully clear of one another; the coefficient of friction could then be vanishingly small.

Those positive charges collectively act as *low-friction bearings*. They keep the respective surfaces from interfering with one another, much like magnetic repulsive forces keep the Shanghai mag-lev train from scraping on the rails beneath. The repulsive forces here come from incident radiant energy: so long as radiant energy is available for sustaining EZs and keeping charges separated, hydronium ions will populate the gap and friction will remain low. Hence, water-based lubrication comes cheaper than oil — practically free, if you can properly exploit the sun's energy.

Danger lurks, however, if those hydronium ions escape. Since hydronium ions repel one another, their escape from the narrow gap is possible — even likely. If enough hydronium ions escape, then the few remaining positive ions can glue the negative EZs together, forming a crystal structure running all the way from one material surface to the other. Such a continuous structure could effectively lock those materials together.

This kind of adherence may explain the common glass-slide phenomenon. Glass slides stacked face to face separate easily if they are dry. However, if a thin film of water gets between any two slides, then pulling them apart can require a team of horses (although sliding remains relatively easy). The major effort needed to pull them apart may reflect the formation of a continuous EZ running from one slide to the other and the substantial forces (Chapter 4) holding those EZ planes together.

On the other hand, those slides could separate easily if enough hydronium ions managed to lodge midway between the respective EZs. Arranging that is not easy with the glass-slide-sandwich configuration: the open edges allow those ions to escape. Thus, maintaining low friction requires keeping those hydronium ions in place, and aqueous low-friction systems do just that.

In fact, water-based lubrication is a far-from-exotic phenomenon; it shows up in everyday situations. For example, wet logs are slippery, and wet floors are more hazardous than dry floors. Water lubrication was the accepted standard before the advent of petroleum-based lubricants. It is now making a comeback. An internet search for "water hydraulics" will yield an array of applications, particularly in the area of food processing, where even minor oil contamination has intolerable consequences; on the other hand, a bit of water in the food won't matter very much.

Modern work with polymers amply demonstrates the effectiveness of water lubrication. Common substances sliding past one another ordinarily have frictional coefficients on the order of 1. When hydrated, however, those polymeric surfaces can have frictional coefficients as low as 0.00001.[1] Hydration can reduce friction by some hundred thousand times.

The reasons for water's extraordinary lubricational efficacy have remained obscure. We can now infer that the responsibility lies with its hydronium ions. By repelling one another, those positive ions push apart the asperities that ordinarily impede shear. Hence, the surfaces can glide past one another practically without friction.

Why Your Joints Don't Squeak

Bones press upon one another at joints. The bones may also rotate, as during deep-knee bends and push-ups. You'd think that rotation under pressure might elicit squeaky frictional resistance, but joint friction remains remarkably modest. Why?

The ends of bones are lined with cartilage. Those cartilaginous materials do the actual pressing. Hence, the issue of joint friction reduces to the issue of the cartilaginous surfaces and the synovial fluid lying in between. How does this system behave under pressure?

Cartilage is made of classic gel materials: highly charged polymers and water; in other words, cartilage is a gel. Since gel surfaces bear exclusion zones, cartilage surfaces should likewise bear EZs, creating many hydronium ions in the synovial fluid between. The fluid itself may contribute additional hydronium ions, from the EZs of the molecules lying within that fluid. Thus, where two cartilaginous surfaces lie across from one another, many hydronium ions will lodge between them. The hydronium ions' repulsive force should keep the cartilage surfaces apart — some scientists maintain that those surfaces never touch. That separation ought to account for the low friction.

For such a mechanism of repulsion to actually work, some kind of built-in restraint should be present to keep the hydronium ions in place; otherwise, charge loss would compromise lubrication. Nature provides that safety net: a structure known as the joint capsule envelops the joint. By constraining the dispersal of hydronium ions, that encapsulation maintains low friction. That's why your joints don't ordinarily squeak.

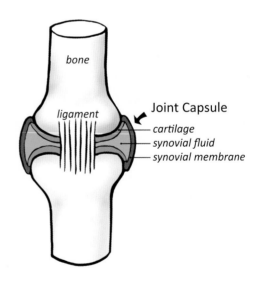

Enveloping the joint, the capsule ensures that fluid hydronium ions don't disperse. The concentrated hydronium ions assure low friction

(ii) Prying Surfaces Apart

This "proton push" can not only reduce friction, but also wedge surfaces apart.

Consider the pyramids. Pyramid construction required finishing blocks cut from the monumental slabs of granite ordinarily found in quarries. To make the necessary cuts, the Egyptians employed a trick: they would pound wooden wedges into cracks, and then hydrate those wedges (**Fig. 12.3**). The Egyptian sun provided plenty of radiant energy for drawing the water into the wood (Chapter 11). The growing EZs would then release protons. The pressure produced by the growing EZs, and especially by the released protons, was evidently sufficient to crack open the solid rock.

If you doubt the verity of this ancient Egyptian feat, consider what can happen when ground water swells tree roots beneath city sidewalks. I recall the ginkgo trees that my neighbor and I planted to grace the sidewalk in front of our English Tudor homes. Tom and I viewed the trees as esthetic assets. The Seattle Engineering Department did not, and ordered their immediate removal. Ginkgo roots, they explained, can exert enough pressure to break the sidewalk. Having witnessed such breakage in Brooklyn's heavily concreted neighborhoods as I grew up, I could not rightfully argue with the authorities — although I do think wistfully of how beautiful those ginkgos might have been if allowed to develop to maturity.

Another example of proton pressure can be found in the simple nut. Nuts are embryonic plants that must break out of their shells by imbibing water. The task is formidable: according to standard plant biology sources,[2] breaking a walnut shell requires a pressure of 600 pounds per square inch — approximately the pressure exerted by three husky men resting their collective weight on a postage stamp.

From that magnitude of pressure, you can better appreciate the truth of various legends: granite boulders split open by oak seedlings lying in cracks; metal containers deformed by dried lentils inside, absorbing moisture from the air; and sailing ships split apart as water seeped into their rice-carrying holds. In all of those scenarios, the likely culprits are those

Fig. 12.3 *Cracking rocks by exploiting hydration.*

pesky protons. So long as radiant energy drives their release, those protons will inevitably exert pressure, which can build to astonishing magnitudes.

(iii) Making Ice Slippery

Protons may also explain the slipperiness of ice.

You may think of ice as being totally solid. In fact, a thin film of water lies on its surface. Michael Faraday first suggested this film's presence in 1842, opining that the water might explain ice's slippery nature. Since then, many experiments have confirmed the film's presence. If your ice skates press down on liquid water instead of solid ice, then you might anticipate easier skating.

An abundance of protons in that water layer would reduce friction even more. Protons do seem to be present: ice, as you may remember from Chapter 4, comprises EZ planes glued together by protons. Melted ice should therefore contain the constituent parts: EZ plus protons; the EZ's presence in just-melted ice has been confirmed.[3] Hence, protons should be present as well. Those protons should help reduce friction (**Fig. 12.4**).

However, that is still not the full story — or even the main story. Many additional protons are created by skate pressure. The skater's entire weight rests on two narrow blades, and more often on only one blade. This creates a pressure equivalent to standing on the floor with several elephants piled on your shoulders.

Applying extreme pressure may compress the ice, squeezing out the interplanar protons (**Fig. 12.5**, *panels a, b*). Those protons will then lie on the ice surface, leaving proton-free ice beneath. (That proton-free ice is the same as the EZ; see *panel c.*) The higher the pressure, the larger the number of squeezed-out protons — a convenient feature if you want to maintain slipperiness no matter how hard you bear down.

On the other hand, if those surface protons are lost through evaporation, then said slipperiness can disappear with the protons. Think of an ice cube sitting in dry air for some time. If you grab the cube, your fingers stick. If the protons are lost evaporatively, then the ice

Fig. 12.4 *Protons separate skate blade from hard ice. The repulsive charge confers low friction.*

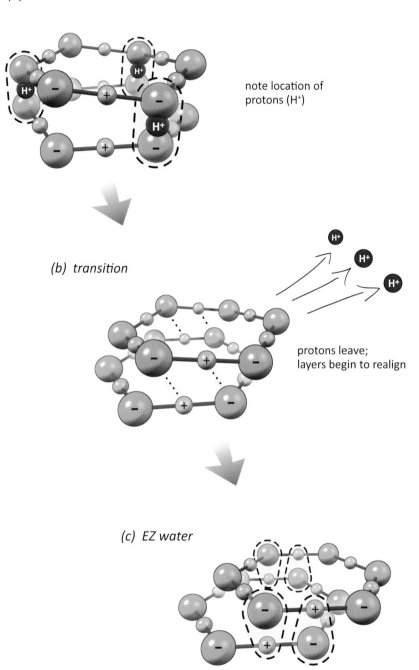

(a) ice

note location of
protons (H+)

(b) transition

protons leave;
layers begin to realign

(c) EZ water

Fig. 12.5 *The effect of pressure
on ice. The pressure squeezes out
protons, which converts the ice to
EZ water.*

Fig. 12.6 *The unfortunate result of a low-temperature misadventure.*

surface will bear the negatively charged EZ alone; that negative charge will induce an equal and opposite charge on any nearby surface — like your fingers. So your fingers stick to the ice. Further, if the water in your skin goes on to freeze, then you and the ice may become more permanently united — like a kid who dares touch a frigid lamppost with his tongue (**Fig. 12.6**).

This sticky scenario reminds me of a story I heard from a young Seattle couple on a blind date. They planned a winter drive to the nearby Cascade Mountains for some casual drinks amidst amiable skiers. Everything went well, but the drive home took longer than anticipated, and nature began to make its inevitable call. When the call became urgent, the gentleman had no choice but to pull off to the side of the road to allow his new friend to do what had to be done. Gallantly, he averted his eyes as she lowered her pants, balanced herself against the frozen car, and proceeded to relieve herself. It was a welcome relief — until she realized that her rear end had become frozen onto the side of the car.

If that story brought a chuckle, you might also appreciate how the story resolved. The distressed woman obviously couldn't just stay put for the night. To gain freedom from her unseemly bondage, the damsel

More Perils of Icebergs

Ships passing through regions of melting icebergs sometimes lose their ability to move swiftly. In some instances, the ships may even get forced to a halt.[w1] Such resistance doesn't necessarily come from Titanic-like hits. Rather, the freshly melted water remains thick and viscous. Plausibly, this high viscosity may derive from massive concentrations of EZ water accumulating from the just-melted ice. Ships passing through this medium may suffer impeded progress.

required a quick thaw, which in turn required a source of warm fluid. Can you imagine where the gentleman got that warm fluid?

The main point is that ice surfaces can exhibit two extremes: very sticky or very slick. The distinction lies in surface protons: their absence permits substances to stick because of electrostatic adhesion; their presence creates a repulsive force which pushes the surfaces apart, yielding the ultra-low friction that permits skates to practically fly over ice.

(iv) Running Batteries

Let us turn from the baseness of urinary function to the more lofty realms of energy, and specifically to the battery. Batteries deliver electrical energy. Modern batteries can do so prodigiously, and yet the lowly potato battery can provide much the same function, albeit in smaller amounts: merely stick two dissimilar metal electrodes into a large potato (or two smaller ones), and voila! — enough energy to power a digital clock (**Fig. 12.7**).

That such low-tech materials can accomplish much of what today's high-tech batteries cans accomplish may seem surprising. Of course, the low-tech versions generate far less power, and they are far less durable. Yet they generate energy by employing similar components: dissimilar electrodes, one being a highly reactive metal such as zinc.

Even Volta's original battery used comparable components. Built more than two centuries ago, Volta's battery comprised discs of zinc and copper (later zinc and silver) separated from one another by brine-soaked cardboard or cloth. The metal pairs were then stacked to form a pile, appropriately called the Voltaic pile (**Fig. 12.8**).

Thus, similar features appear in batteries ranging from Volta's original device to the potato battery to modern day alkaline batteries. One electrode is made of zinc or some other reactive metal, while the other is made of a less reactive substance. In between the electrodes lies some ion-containing medium that can sustain current flow. It is widely thought that energy production derives from electrochemical reactions originating at the metal interfaces. Those reactions supposedly push

Fig. 12.7 *The potato battery. Tomatoes work equally well, and if vegetables are out of season, then salt water can do the trick.*

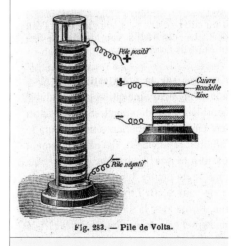

Fig. 283. — Pile de Volta.

Fig. 12.8 *Original pile developed by Volta.*

charges through the ion-containing medium and then out the terminals, providing the output energy.

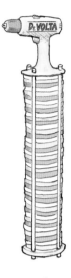

While reactions certainly take place at those interfaces, the phenomena considered earlier in this book may tell us something about the nature of those reactions. Again and again, we have seen absorbed electromagnetic energy separating charge via the EZ mechanism. Batteries also separate charge. The question arises whether the batteries' charge separation might involve EZ-based charge separation.

early cordless drill concept

This question rose in importance when it became clear just how much energy common batteries deliver. Consider the alkaline battery. Alkaline batteries have been used in one form or another since Thomas Edison invented them over a century ago. Over its lifetime, a modern AA alkaline battery can deliver as much as 1 mA of current for 1,400 hours. The product of current and time yields 5,000 coulombs; that is, an AA battery can deliver 5,000 coulombs of charge. A typical lightning bolt discharges 15 coulombs. Thus, the diminutive AA battery contains enough internal energy, in theory, to drive more than 300 lightning bolts.

Could a chemical factory so diminutive contain that much energy? Perhaps. But some of that energy might come from elsewhere — i.e., from the absorbed electromagnetic energy mechanisms described earlier. The reaction products might be the same as those envisioned conventionally; but the energy driving those reactions would come from electromagnetic energy absorbed from outside. Visible light can't pass through battery casings; however, infrared energy can. Infrared energy absorbed by those casings gets reradiated into the battery's interstices, and the question arises whether such energy might contribute to the battery's remarkable output.

Curious to see whether EZ-like features appeared next to those reactive metals, we stuck a piece of zinc into water. With the addition

OUT-ON-A-LIMB METER

of microspheres, we immediately saw large exclusion zones developing from the zinc surface (**Fig. 12.9a**). The EZs grew to approximately 200 μm. Other reactive metals exhibited similar features. It seemed plausible that the reactive metal surfaces common to so many batteries could separate charges in much the same way as do other EZs.

(a)

(b)

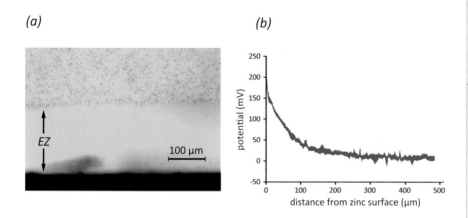

Fig. 12.9 *Zinc-surface features.*[4]
(a) *Exclusion zone found next to zinc.*
(b) *Electrical potential within the EZ*

We tested to see whether charges were actually separated. Using the same microelectrode apparatus employed earlier, we struck gold: we found that the EZs were indeed charged. The EZs next to reactive metals were positively charged (**Fig. 12.9b**); pH-sensitive dyes confirmed that the water beyond those EZs contained free negative charges, presumably in the form of OH⁻ groups.[4]

Thus, charges really were separated next to reactive metal surfaces, although the polarity was the less common one. We were able to draw substantial currents from those separated charges,[5] much like the currents drawn from other EZ systems. This result seemed pivotal: since batteries use reactive metals, and since reactive metals exhibit EZ-based charge separation, the result implied that at least *some* of the battery's energy could come from external energy input.

The question lingered: *how much* of that energy? We think of modern batteries as packing all of their deliverable energy inside — a self-enclosed storehouse of chemical energy. Considering the batteries' phenomenal lifetime output, one wonders whether this can be true. Batteries certainly absorb electromagnetic energy, but do they make use of that bounty, or discard it? Plausibly, the diminutive chemical factory that we call a battery might produce its electrical output by using at least some of that incident electromagnetic energy (**Fig. 12.10**).

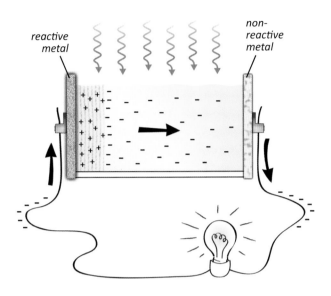

Fig. 12.10 *Possible EZ-based production of electrical energy in standard batteries. Absorbed electromagnetic energy separates charge and thereby contributes to the current.*

(v) Protons Drive Catalysis

Catalysis is the seemingly mysterious process that accelerates chemical reactions — sometimes by millions of times. The catalyst is not consumed; it remains in place and accomplishes its wondrous feats again and again.

Most common among catalysts are the so-called acid catalysts; they accelerate reactions by mobilizing protons. Other catalysts accelerate reactions by mobilizing OH⁻ groups. I wondered whether those charge groups might come from EZs (**Fig. 12.11**). Exclusion zones commonly produce free H^+ and, less commonly, free OH^-. Thus, hydrophilic surfaces ought to be natural catalysts. Those with the highest charge should be the most catalytic. Nafion's potent catalytic activity would fit that expectation.

EZ-based catalysis would not come free. Radiant energy builds EZs, which in turn generate the required charge groups. So EZ-based catalysis can yield something potentially useful, but you need to pay for it with energy.

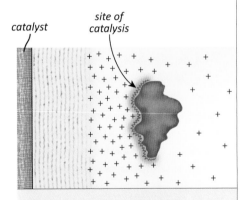

Fig. 12.11 *EZ-based catalysis. The catalyst surface nucleates EZ growth, which results in proton (or OH⁻) release. The protons (or OH⁻ groups) accelerate the target reaction.*

Starting Your Car on a Cold Morning

The weather outside is frigid. Because of the cold, the oil in your car's engine has the consistency of peanut butter. Pistons struggle against that viscous glop as you begin cranking the starter motor, but to little avail. The car won't start. Continued cranking only drains the battery, and before long it is practically dead. Time to give up and call a tow truck.

A different strategy might save you the frustration. If you stop cranking and patiently wait before trying again, then the engine may quickly catch and you can go about your business. The mere act of waiting seems magically to recharge your battery.

Why should waiting make a difference? Auto batteries contain water and acid, a combination that builds EZs (see **Fig. 10.9**). If EZ energy contributes to the battery's energy, then getting energy from your battery demands maintenance of the EZ charge. You can rebuild EZ charge in at least two ways:

from the tow-truck's portable electrical charger; or by absorption of the radiant energy coming from the slightly warmed engine. The latter approach requires time. That's why waiting can make a difference.

With the EZ charged from radiant energy absorption, and the peanut butter melted from earlier cranking, your car may finally start. Patience may be virtuous; here, it can also save you the expense of a tow truck.

If incident radiant energy ultimately drives catalysis, then we would expect that light might be a recognized driver of at least some catalytic processes. That is indeed the case. A well-known example, titanium dioxide, exhibits a spectacular rise in catalytic activity when exposed to UV light.

To check whether that particular feature is EZ-based, we examined sheets of titanium dioxide immersed in water. With no incident light (beyond the modest amount used for visualization), no EZ was present; when UV light illuminated the sheet prior to its placement in water, subsequent immersion showed exclusion zones that grew to approximately 200 μm. Using pH-sensitive dyes, we confirmed the presence of numerous protons in the bulk water beyond the EZ. Thus, EZ-based protons were in place and ready to perform their catalytic task. Ultimately, incident light powered those catalytic protons.

Platinum, another common catalyst, initially failed to exhibit clear exclusion zones, although we could find some near-surface zones with reduced microsphere concentrations.[4] Subsequent studies showed light-dependent changes in the water near platinum's surface,[6] as well as clear EZs when the platinum was electrically connected to a reactive metal.[5] Hence, EZ-based charges might explain platinum catalysis as well, although this possibility needs further study.

Catalysts also present themselves in biological contexts, in the form of enzymes. Enzymes are large proteins that accelerate biological reactions. The catalytic mechanism is presumed to reside in enzyme-specific interactions with reactant molecules. Up to the early 20th century, however, the prevailing view differed: it was supposed that enzymes induced changes in the surrounding water, which then accelerated reactions in molecules nearby. Enzyme surfaces commonly bear substantial negative charge (as do most protein surfaces); hence, those surfaces should contain EZ layers. If so, then, biological catalysis may resemble generic catalysis, involving little more than high concentrations of EZ-generated protons.

(vi) Protons Power Fluid Flow

Finally, consider how those versatile protons might generate flow.

Envision a corked bottle of water immersed in a large water bath (**Fig. 12.12**). In a thought experiment, suppose that you had some way of infusing protons into that corked bottle. The ions would repel one another, creating pressure inside. If you then pop the cork, those confined hydronium ions would quickly escape — like air from an over-inflated balloon. The protonated water would gush outward, creating a measurable flow. That's the kind of flow of which we speak.

Flow of this nature could persist indefinitely if the protons and water were continually replenished. If EZs are involved, then proton

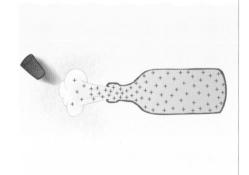

Fig. 12.12 *Possible EZ-based production of electrical energy in standard batteries. Absorbed electromagnetic energy separates charge and thereby contributes to the current.*

216

replenishment is natural, for EZs generate protons continuously so long as ambient energy remains available to drive their release. The protons immediately form hydronium ions. Those charged water molecules will then move toward regions of lower charge. Hence, *sustained water flow occurs inevitably in almost any scenario involving EZs and radiant energy.*

Experiments have confirmed this inevitability: we have seen sustained water flows in a variety of physical configurations in which EZs are present. You've already read about some of them (**Chapter 7**), but a collective recap is worthwhile.

• The first configuration is the hydrophilic tube (**Fig. 12.13**), just inside of which an annular EZ ring forms, generating protons in the tube's core; those core protons form a hydronium ion gradient from the core to the tube's openings, which appears to drive the flow one way or the other.

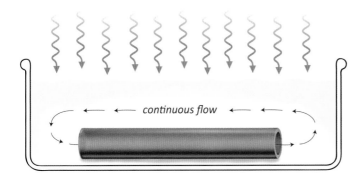

Fig. 12.13 *"Spontaneous" flow through hydrophilic tubes.*

• A second configuration resembles the first, but with a hole in the tube's wall. When a small hole is punched in a hydrophilic tube submersed in water, the water consistently flows inward through the hole (see **Fig 7.11**). Again, hydronium ion gradients seem responsible for driving this flow.[7]

• A third configuration showing charge-driven flow occurs with gel beads (see **Fig. 7.12**). Sitting on the floor of a water bath, a hydrophilic bead becomes enveloped by an EZ shell. Hydronium ions lie beyond. Those hydronium ions consistently flow downward, as diagrammed (**Fig. 12.14**). The driving force seems to be a vertical hydronium ion gradient, which might arise because the top of the suspension receives more radiant energy than the bottom. The downward flow then draws top-level water from all directions toward the bead, replacing the molecules lost from flowing downward. Once the downward-flowing molecules reach the bottom, they have no choice but to flow away from the bead. So the water circulates, driven by the vertical hydronium ion gradient.

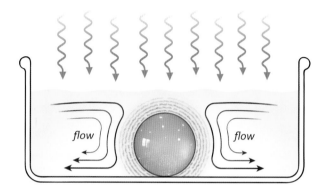

Fig. 12.14 *Sustained flow pathways around a hydrophilic sphere.*

Thus, charge-driven flows occur in a variety of configurations. Most any hydrophilic surface immersed in water is sufficient. The resulting EZs generate protons; the inevitability of hydronium ion gradients makes flow similarly inevitable.

In fact, some of the flows that scientists reflexively ascribe to thermally induced density gradients could easily arise from charge gradients. Charge-based forces are *much, much* greater than gravitation-based forces. To appreciate the difference, think of a proton and an electron situated near one another. Ask which force dominates: electrostatic attraction or mass-based gravitational attraction? Surely you'd say that the electrostatic force dominates, but the surprise is the magnitude: a factor of 10^{38}. Thus, charge-based forces can be overwhelmingly dominant. As we have just seen (and will see again when we deal with warm water in Chapter 15), small charge gradients can generate major flows.

Because all of these flows are charge driven, you might opt to classify them as osmotic (Chapter 11). Hydronium ion gradients constitute a powerful and universal force of nature.

Summary and Reflections

This chapter has focused mainly on protonated water — the positively charged water molecules that inevitably arise from the presence of exclusion zones. Those charged molecules can wreak havoc. Their sundry actions include reducing friction, wedging surfaces apart, making ice slippery, running batteries, driving catalysis, and powering fluid flows. A good case can be made that all of these processes arise, at in least part, from the charges generated by EZ formation.

I encourage you to contrast this view with the prevailing views on the origin of each of these phenomena. Students entering my laboratory invariably find themselves eager to check prevailing views in order to familiarize themselves with the territory. Many of them return confused — bewildered by interpretational complexities and feeling inept because of their inability to fathom what they presume others can understand easily. Perhaps you'll fare better than those hapless students.

The central message of this chapter is that all of these phenomena may arise as direct outcomes of EZ generation. Radiant energy drives each of them. Plants certainly exploit radiant energy, but the idea that these diverse aqueous systems may also exploit radiant energy has remained outside the purview of current thinking, and I believe that this deficit explains why investigations have not progressed to clear end points. If an EZ is present, then we must assess its impact, and that is what we have attempted to do.

EZ charges have dominated our thinking over many chapters, and we're not yet done. Next come bubbles and droplets. You've seen bubbles forming in boiling water, but did you ever stop to think about how they form? The next two chapters address that question and show that EZ-based charges once again play a central role. Bubbles may seem mysterious — almost romantic; if the forthcoming chapters succeed in demystifying something romantic, I offer my apologies in advance.

13 Droplets and Bubbles: Siblings in the Water Family

M̲ost passengers were asleep. I was on a transatlantic flight, busy as usual with my laptop as flight attendants made their rounds, dutifully refilling everyone's cups with water. The attendants seemed alert to the risks of dehydration. I nodded each time in appreciation, drank a few sips, and finally drifted blissfully into dreamland.

When I awakened a few hours later, I noticed that my half-filled plastic cup looked different. Bubbles clung to the inside walls (**Fig. 13.1**). I wondered, how did those bubbles get there?

Surely a good deal of air must have dissolved in the water during my snooze; that dissolved air ought to have sufficed to create the bubbles. Right?

That explanation seemed adequate until I wondered further: the bubbles I saw were on the scale of *milli*meters while the gas molecules filling the bubbles are the size of *nano*meters. To build a bubble whose volume is 10^{18} greater than its constituent gas molecules, you need those molecules by the gazillion. The obvious questions: What directs those many gas molecules toward the sites of growing bubbles, as opposed to sites in between? Why only a few isolated bubbles instead of, say, one huge bubble?

Bewildered and still half asleep, I began to fantasize. I imagined myself shrinking to the size of a proton and looked around. I had no trouble visualizing the dissolved gas molecules around me; but I did have trouble envisioning what might direct those molecules *preferentially* toward the growing bubble nearby. And if one of those gas molecules could somehow manage to approach that friendly bubble nearby, how could it penetrate the tensioned water-gas interface to get inside? Wouldn't that break the bubble?

Continuing my underwater odyssey, I glanced upward, just above the water surface. There I could see additional bubbles clinging to the

Fig. 13.1 *Bubbles eventually gather on the walls of plastic cups filled with water.*

side of the cup. I quickly reminded myself that I was probably mistaken: above the water's surface, usually droplets clung, not bubbles. Bubbles and droplets can sometimes look awfully similar from the outside; but bubbles contain gas, while droplets contain liquid. How might I know which one I was seeing?

When I returned from my waking reverie, I came to appreciate that nobody had easy answers to those questions. Even Einstein, famous for his "thought experiments," might not have figured them out.

For the last question, at least ("How can I tell a bubble from a droplet?"), you'd think that location would provide a helpful clue: bubbles can exist underwater, but droplets should not — how could water droplets exist within a body of water?

Paradoxically, droplets *can* exist within water, and I'll offer evidence in a moment. Those underwater droplets turn out to be key. Their features will lead the way toward answering even the most vexing of the above questions: how can a growing bubble fill itself with gas?

Bubbles and Droplets: Siblings in the Vesicle Family

It all started at weekly laboratory meetings. One member of the lab kept stumbling: he said droplet when he meant bubble (and bubble when he meant droplet) so frequently that his error became a common vehicle for teasing: how do you know a bubble from a droplet?

The confusion first arose as we looked into the "bubbles" that often form underwater near the Nafion-water interface. We would frequently see several of them develop in the course of an experiment. Because they formed underwater, we reflexively assumed they must be bubbles.

We found, however, that those putative bubbles behaved more like droplets. When penetrated by a sharp probe, more often than not nothing happened. Punctures instantly break tensioned balloons, but here we saw no breakage. Further, extreme pulling or squeezing had a surprisingly limited impact; the battered entities were cohesive enough to quickly regain their roughly spherical shape. These "bubbles" behaved differently from any bubble we'd seen.

To check more carefully, we set up a guillotine (**Fig. 13.2**). The submerged blob was gently placed on a concave Teflon surface, where it lodged securely. Then we sliced the "bubble" with a sharp knife. As the knife passed through — all the way through so that the body was dismembered into two distinct parts — the putative bubble would return to its initial configuration: a single intact blob. The underwater blob seemed immune to decapitation.

In no way did this entity behave like an ordinary bubble. Piercing, stretching, and squeezing produced only transient effects; the ravages of the guillotine had no enduring impact. What seemed assuredly to be a bubble — for it was under water — behaved more like a sticky, gel-like blob or like some kind of cohesive droplet.

This unexpected result taught us a lesson: without proper testing, it is not always possible to distinguish bubbles from droplets; reality may turn out contrary to what you expect. We've just seen droplets underwater. And bubbles resting on water surfaces sometimes turn out to be droplets: water *droplets* can persist on water surfaces for extended times without dissolution (see **Fig. 1.5**). If you'd not seen those persisting water droplets before, you'd probably think they were bubbles — but they're actually droplets. Also, consider the blobs in **Figure 13.3**. Are they droplets or bubbles?

The potential for confusion seems clear. Bubbles resemble droplets, and droplets resemble bubbles. You might say they are look-alike members of the same family. Because of their similarity, we began adopting a generic nomenclature — we refer to both entities as "vesicles." We define "vesicles" as spherical entities whose insides may contain either liquid or gas. Later, we will provide evidence that the liquid inside may *transition* into gas.

For the moment, the point is that droplets and bubbles bear a striking resemblance to one another.

How Are Bubbles and Droplets Similar?

One evident common feature is the enveloping membrane. In bubbles, membranes can be plainly seen, for example during boiling.

Fig. 13.2 *Bubble guillotine schematic. Attempts to slice submerged "bubbles" often proved futile; the bubble re-formed.*

Fig. 13.3 *Water on a hot skillet.*

Fig. 13.4 *Droplet shape. The force of the distending pressure balances the inward force of membrane tension, promoting a spherical shape.*

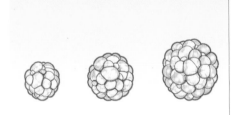

Fig. 13.5 *Droplet growth by fusion. Multiple mini-droplets may coalesce to form larger droplets.*

In droplets, we infer the membranes' existence because the droplet's near-spherical shape can be explained by an enveloping membrane stressed by internal pressure (**Fig. 13.4**).

In the absence of a tensioned membrane, bubbles and droplets might assume shapes as irregular as amoebas; tumor-like bumps could protrude everywhere. Thus, scientists accept the presence of some kind of membranous sheath as a given. Pressured sheaths surround all vesicles.

Less clear is the source of the pressure. Scientists routinely lay the blame on the sheath's rigidity: they envision a stiffened sheath as building pressure inside the vesicle; however, even sheaths of steel offer no guarantee of pressure within. We cannot say at the outset whether the prime mover is the sheath's stiffness or the bulk material enclosed within the sheath. The pressure's origin remains to be addressed, and we will do so shortly.

At any rate, the droplet's near-spherical shape does imply the presence of some kind of enveloping sheath. That inference seems straightforward enough. What doesn't seem to fit, however, is the guillotine result: if a tensioned membrane sheathes the droplet, then fracturing that membrane should be fatal; the droplet should disintegrate. Why should the guillotined droplet reform as readily as it does?

A plausible explanation may be that some droplets comprise clusters of smaller droplets (**Fig. 13.5**). The mini-droplets would cluster together just as bubbles do; they could subsequently recluster following mechanical separation. Indeed, underwater droplets sometimes looked like clusters in the microscope, although we have yet to capture sufficiently detailed images to be certain.

The cluster idea gained traction from high-speed video recordings of falling raindrops. As raindrops descend, many of them break up into myriad mini-droplets.[w1] Those mini-droplets could form *de novo* as they fall, but they could also preexist, separating when facing the forces encountered during their descent. If so, then droplets more generally could resemble the cluster shown in **Figure 13.5**.

In this context, the guillotine result could make sense. Applied forces could sever the subgroups of droplets; but cohesive forces between their surfaces could then reconnect the constituent subgroups in much the same way that separated bubbles tend to reconnect.

Thus, droplets may behave like bubbles. Bubbles can exist in clusters; cluster members can fuse into a single larger bubble. The same may hold true for droplets: the droplet could comprise a series of mini-droplets; or, those mini-droplets could fuse into one massive unit. All of these fusing events would arise from the properties of the encompassing membrane.

The reason for dwelling on this issue goes beyond understanding how droplets behave in guillotines. We haven't a clue how droplets or bubbles form. We need ideas. One idea follows from our discussion: Bubbles and droplets both contain membranes; if those membranes happened to comprise the same material, then a transition from droplet to bubble (or vice versa) is conceivable. In other words, *bubbles could arise from droplets whose liquid insides have turned to vapor*. Droplets might be the progenitors of bubbles.

To pursue this speculation, we need to consider the relevant evidence. We ask first whether membranes really do surround these two vesicular structures, as implied. And, if so, what might be the membranes' composition? If they turn out to have the same compositions, then we will have arrived at an auspicious launch point for considering the droplet-to-bubble transition.

Droplets Contain EZ Membranes

The first evidence that led us to think seriously about droplet membranes came from the floating droplet experiment (see **Fig. 1.5**). Water droplets falling upon water are widely supposed to instantly coalesce, but we observed that those droplets could persist for many seconds.[1] Something evidently retarded their collapse, and a restrictive membrane seemed a logical candidate.

The prevailing explanation for delayed coalescence had been the presence of an invisible film of air trapped between the falling droplet and the surface. However, we dispelled that notion by observing that droplets could persist even after rolling beyond the putative cushion of trapped air. In fact, rolling *prolonged* the droplets' separate existence.[1] Something other than trapped air was evidently responsible, and we considered a membranous sheath the most likely candidate. The sheath would need to dissolve to permit coalescence, and dissolution could require time.

As for the nature of that sheath, the answer seemed deductively evident. Droplets comprise water alone. Therefore, any membranous sheath had to be built of some form of water. The two plausible options were bulk water and EZ water; EZ water seemed the more natural candidate for explaining the observed persistence.

An EZ sheath also made sense functionally. A shell made of EZ material could contribute protons. If those protons accumulated inside the droplet, then the repulsive forces could generate the pressure needed to explain droplet roundness. That seemed downright elegant. Further, the EZ shell made sense theoretically. An extensive theoretical and experimental study concluded that a negatively charged shell of this sort was a minimum requirement for a droplet's existence.[2] An EZ shell could certainly satisfy that requirement.

Given these hints, we proceeded to check whether we could identify any such EZ shell experimentally. We found three pieces of relevant evidence.

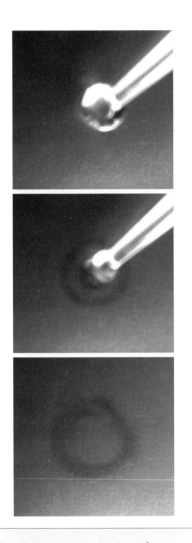

Fig. 13.6 *Microscopic view of droplet dissolution. A ten-microliter droplet of water was released onto water containing 1-μm carboxylate microspheres. A ring-like clear zone implies an EZ sheath surrounding the droplet.*

• The first evidence came from experiments that began with a microsphere suspension. Onto the surface of that cloudy suspension, we released water droplets. We reasoned that an EZ shell might behave differently from a droplet's interior: while the water inside the droplet might pass through a breach in the shell and directly into the cloudy suspension beneath, the droplet's shell might not; an EZ shell might remain on the surface of the suspension. **Figure 13.6** confirms that clear zone. The clear zone appears at the onset of droplet coalescence and expands outward as coalescence proceeds. Its annular shape and clarity seemed consistent with remnants of an EZ shell.

• A second piece of evidence for an EZ shell comes from optical absorption studies. If droplets contain EZs, then they should exhibit the characteristic EZ absorption at 270 nm. We therefore collected droplets of rain. We tested rainwater samples in a UV-VIS spectrometer. **Figure 13.7** confirms the presence of the characteristic absorption peak in samples collected on two different days.

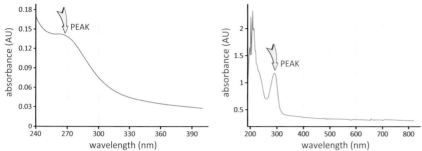

Fig. 13.7 *UV-Vis spectra obtained from raindrops collected from two different rainfalls. A peak near 270 nm is evident on the characteristic descending curves.*

• EZ membranes are also implied by a third observation: droplets dried on glass slides. My colleague Georg Schröcker studies those droplets as a hobby. Dried droplets of water from various sources leave ring-like residues (**Fig. 13.8**). Typically, the central zones are empty, which means the water has fully evaporated. The annular zone surrounding that central zone, however, does not evaporate; a residue clings to the hydrophilic glass surface — the kind of residue one might expect of an adherent EZ. Additionally, the annulus fluoresces mainly blue-violet, much the same as the fluorescence of the EZ.[3] Further, the residue sometimes appears layered like an EZ, although the scale is larger than molecular. Hence, the shells of dried droplets have much the same features expected of EZ shells.

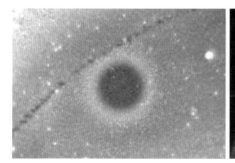

This set of results supports the proposition that droplets possess an EZ shell. They do not prove the point but are strongly suggestive. If the protons associated with an EZ were to lodge within the confines of such a shell, then the droplets' spherical shape would be explained as well, because proton repulsions create pressure (Chapter 12).

Fig. 13.8 *Dried droplets, viewed in dark-field illumination. **Left**: spring water; **middle**: homeopathic preparation; **right**: dew droplet. Note blue color and concentric layering of shell.*

EZ Membranes Surround Bubbles

We next ask whether EZ shells also surround bubbles. Bubbles contain some kind of membranous envelope, showing up as a hemispherical cap as water boils. What remains unclear is whether those envelopes comprise EZ material.

To check, we tested again for the 270-nm absorption peak. We slowly heated pans of water to temperatures just shy of boiling. Small bubbles formed at the bottom, but most of those popped inside the water. Testing the water spectroscopically to see whether the popped shells' remnants contained EZ material confirmed their presence: all samples showed a 270 nm absorption peak (**Fig. 13.9**).

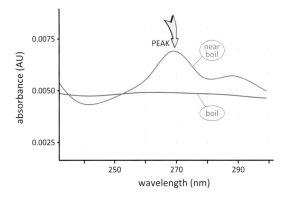

Fig. 13.9 *Absorption spectra of water previously heated to just below boiling and allowed to cool to 60 °C (blue). Red curve was obtained similarly, but water was first heated to boiling.*

On the other hand, samples brought to full boil so that bubbles broke at the top, contained no such peak (red curve, **Fig. 13.9**). Nor could the peak be found in water that had not undergone heating. Thus, bubble shells appear to contain EZ material.

To pursue the EZ-shell question in an altogether different way, we looked at bubble-bubble interactions. If bubbles contain EZ shells, then the bubble-bubble interactions should resemble particle-particle interactions, for both would contain EZ shells. Back in Chapter 8, we considered particle-particle "sociology": like-likes-like attractions drew like-charged particles together into ordered arrays and sometimes into coalescence. We reasoned that EZ-shelled bubbles ought to do much the same. Bubbles should attract each other, order, and perhaps coalesce.

Bubble-bubble attraction is commonly visible in just-poured hot coffee or hot water. The bubbles tend to cluster, leaving bubble-free regions in between. Sometimes smaller clusters will grow into larger ones. Bubble-bubble attractions can also be seen underwater. Those attractions are usually studied in tall water columns, where bubbles released from the bottom can be tracked over long distances as they rise. During their ascent, the bubbles draw together, some coalescing into larger bubbles. This attraction phenomenon has been well studied, although the reason for the attraction has seemed elusive.[4]

Given enough time, like-likes-like attractions between bubbles can bring order. An example is shown in **Figure 13.10**. The ordering in this figure is not precise, presumably because of the surface bubbles' varying sizes. However, some areas with similar-sized bubbles exhibit regular spacing.

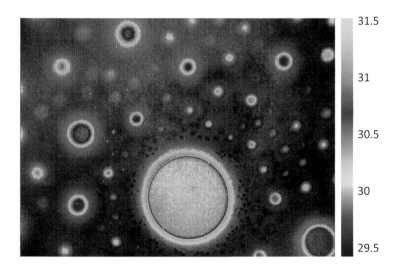

Fig. 13.10 *Surface bubbles tend towards ordered arrangements. Bubbles were created by pumping air into an aqueous detergent (TWEEN 20) solution. Scale in °C.*

In this framework, agents that inhibit the like-likes-like attraction should inhibit clustering. Take, for example, salt. Salt diminishes EZ size,[5] thereby reducing the number of separated charges. Hence, salt should diminish the like-likes-like attraction. Adding enough salt should prevent the bubbles from fusing at all — and that's what experiments have shown: in the presence of salt, small bubbles released at the bottom of a flask no longer coalesce into larger bubbles;[6] they remain separated.

Perhaps the most idiosyncratic expectation of an EZ shell is its attraction to light: EZ-enveloped particles move consistently toward light (Chapter 9). Bubbles behave similarly. Bubbles move toward light both at infrared wavelengths[7] and at visible wavelengths.[8] Researchers ascribe this attraction to a "thermocapillary migration" effect, although they still have not pinned down the basis for that effect. It seems apparent that *light attracts bubbles and particles alike*. If EZ-based charge separation bears responsibility for the particle attraction (Chapter 9), then it plausibly bears responsibility for the bubble attraction. In other words, bubble attraction implies that bubbles should contain EZ shells.

The thickness of a bubble's EZ shell can prove difficult to estimate. Many types of bubbles show membranous caps as they penetrate the liquid's surface. Commonly, the caps are hundreds of nanometers thick, but it is not uncommon to find thicknesses on the order of 1,000 nanometers.[9] A 1,000-nanometer-thick EZ film would contain some 40,000 molecular layers. Even fewer layers could be sufficient to constitute a fairly robust membrane.

Finally, it's worth mentioning that EZ shells naturally accommodate volume changes. EZs consist of layered sheets. With enough distending pressure, those sheets can shear past one another, allowing the shell to expand (and consequently thin). Thus EZ shells can expand with impunity against the force of considerable distending pressure. On the other hand, excessive thinning may bring fracture, and that may be what happens during boiling.

Like droplets then, bubbles seem to possess EZ shells. The EZ's characteristic absorption at 270 nm is confirmed; bubble-bubble "sociology" fits EZ-based expectations; light draws bubbles like it draws EZ-shelled particles; and the EZ's sheet-like structure naturally accommodates volume changes.

So we continue thinking that bubbles and droplets are structurally similar — although the core of the droplet may contain liquid while the core of the bubble may contain gas. This structural similarity will prove consequential as we go on to explore how bubbles form.

Summary

Droplets and bubbles resemble one another. Both entities are characteristically spherical and transparent; and both can exist above or below the water's surface. We speculated that these common attributes might arise from the presence of an enveloping sheath. We found evidence for a membranous sheath in both types of vesicle and found that those sheaths comprised EZ material.

Since EZ buildup generates protons, protons residing inside the vesicle and repelling one another could generate the pressure required for maintaining the vesicles' spherical shape. Those protons could exist in the droplet's liquid and the bubble's vapor alike.

Proton buildup requires external energy. External energy elevation should drive a pressure increase and a resultant vesicle expansion. That expansion can bring consequences — as we'll see in the next chapter. Would you believe a liquid-to-vapor conversion?

"today, Drop, you will become a Bubble"

14 Birth of a Bubble: A Passage to Maturity

Seattle winters often bring gloom. Grey clouds cover the sky, dumping their contents often enough that old-timers seem to have acquired webbed feet. Welcome respites come when the sun pokes through. When those sunny winter periods finally do arrive, nighttime temperatures will often dip down close to freezing.

As I approach my car on those cold winter mornings I often note a fine mist accumulated on the side window. The mist looks like frost, but it's made of fine liquid droplets that wipe free with a cloth. Oddly, the mist appears on the driver's side only and not on the passenger's side. That difference puzzled me for years.

The passenger's side faces my neighbor's home, whereas the driver's side faces relatively open air. I'd understood that facing such a frigid expanse of empty space meant losing heat, which could cause condensation. Nevertheless, I couldn't understand several aspects of that phenomenon: why the condensation appeared as droplets; why those droplets clung so tenaciously to the window; and, most of all, why the droplets consistently failed to form on the side facing my neighbor's house. Was it perhaps my neighbor's radiant charm?

This chapter deals with those droplets, but more generally it deals with vesicles: i.e., droplets and bubbles. It begins where the previous chapter left off: the similarity of droplets to bubbles. Building on that, we will consider whether the similarity has functional significance. I will argue that *droplets are the progenitors of bubbles*.

Questions that will arise along the way include the following:

- How could droplets possibly build inside a bath of water?

- How do multiple droplets merge to create larger droplets?

- How could those larger droplets transition into vapor-containing bubbles?

- How can the coalescence of multiple bubbles lead to boiling?

The chapter will also reach beyond those fundamentals, dealing with related everyday issues: Why do kettles make loud clattering noises as the water inside heats up? Why does that clatter give way to the familiar teakettle whistle? And when you enter a kitchen, why can you smell the soup that's cooking?

The Embryonic Bubble Concept

Bubbles contain gas. In order understand how bubbles form, we must know where that gas (or vapor) comes from.

I recall pondering that question several years ago while traveling on a sleek train from Vienna to Graz. My bubble questions, initially raised during a transatlantic flight (see Chapter 13), still haunted me. What plausible mechanism could draw all of those gas molecules toward distinct spots to form those separated bubbles? And if the gas molecules could manage to get there, then how might they pierce the tensioned bubble membrane without causing the bubble to burst?

Lulled into a reverie by the pastoral scenery, I finally had a eureka moment. Suppose the gas molecules didn't actually *need* to "get there." Suppose the process that created the EZ sheath also created the gas inside? Then those confounding issues might turn moot: the gas molecules might not need to cross any tensioned membrane, nor would they need to find their way to the site of the growing bubble.

While potentially solving both problems, the idea first seemed contrived: What process could create an EZ sheath out of water? And how could that process create the internal gas — presumably water vapor? Those challenges seemed formidable. Yet, the concept seemed promising because it neatly circumvented all the issues that had lent confusion. The concept seemed worth at least putting on the back burner.

Then came a second eureka moment: droplets and bubbles both contain EZ sheaths. Because of this similarity, one entity could conceivably transition into the other. Suppose the droplet formed first. The droplet could form if some hydrophilic surface in the water nucleated an EZ; if that EZ could somehow curl itself into a sphere, then we'd have an EZ-sheathed droplet. Step 1 seemed plausible.

Next came step 2 — the droplet-to-bubble transition: If the droplet absorbed enough radiant energy, then, conceivably, the water inside the droplet might transition into vapor. The droplet would then become a bubble.

By the time my train pulled into the Graz station, I was so intoxicated with this new possibility — almost giddy — that I couldn't wait to find time to further ponder how the droplet could be the embryonic bubble.

Creating the Embryonic Structure

Building an EZ sheath requires an EZ-nucleating (hydrophilic) surface. Nucleators sitting in a cup of water differ in several ways from those we have considered so far. First, the cup of water will inevitably contain dissolved solutes and suspended particles; even "pure" water will contain abundant odds and ends because water is a universal solvent. Those junk molecules remain dissolved/suspended specifically *because* they build EZs around them (see Chapter 8). Therefore, essentially all water contains EZ-nucleation sites.

An additional nucleation site may be the cup itself: if the material is hydrophilic glass, then that surface can create the EZ, especially at local rough spots (see below). If the surface is hydrophobic, then it is theoretically inept; however, substances in the water with net charge may induce opposite charge on the container surface in the standard way, and then cling. The clinging substances serve as asperities, which can then create EZs.

With nucleation sites in abundance, EZs can build in the usual manner, layer by layer. Those layers can also grow laterally: we regularly observe lateral growth in the laboratory when dealing small,

nucleator

Fig. 14.1 *Stages of vesicle formation. **Left:** EZ grows layer by layer, and also laterally (arrows). Protons build positive charge beyond. **Middle:** EZ deflects because of attraction to positive charge. Red arrows indicate direction of deflection. **Right:** continued deflection creates vesicle.*

bounded surfaces immersed in water (**Fig. 14.1**, *left panel*). We find the EZs not only stacking perpendicular to the nucleator surface, but also growing in the lateral direction — often extending beyond the nucleator's corners.

Such EZs generate protons in the usual way, and the protons quickly convert into hydronium ions. Some of those ions disperse because positive charges repel; those hydronium ions are lost. Other hydronium ions remain close to the surface because of their attraction to the negative EZ (see Chapter 5). That attraction creates two responses: positive hydronium ions moving toward the negative EZ, and the negative EZ — thin and flexible in its initial stages — deflecting toward the positive hydronium ions. EZ flanks should thus draw continuously toward the highest concentration of positive charge (**Fig. 14.1**, *middle panel*). We have experimentally confirmed the negative EZ's deflection toward positive hydronium ions (see **Fig. 9.9**).

Such continuous deflection leads inevitably to curvature (**Fig. 14.1**, *right panel*). As hydronium ions build up and lateral growth continues, newly built regions of the EZ will keep deflecting toward the central hydronium ions. The EZs should eventually meet, creating a circular structure. Of course, the real structure is spherical, not the circular shape shown in the figure. Spherical structures should close naturally (**Fig. 4.11**). Those closed spherical structures constitute tiny droplets.

Each such mini-droplet will contain a negatively charged EZ shell, enveloping liquid water and hydronium ions. The hydronium ions repel; as they flee each other in every direction, they exert pressure on the EZ shell, conferring roundness (**Fig. 14.2**). A spherical bubble is not yet born, but a tiny embryonic vesicle has been conceived, ready to begin its development toward maturity.

It is worth pointing out that those interior positive charges should not necessarily equal the EZ's negative charge. Such a balance might be reflexively anticipated; that would confer neutrality. However, some hydronium ions will have inevitably escaped during vesicle buildup because of repulsive dispersal (**Fig. 14.1**). Hence, the vesicle is *not* neutral. *The closed vesicle should bear a net negative charge.* That negativity will turn up again later.

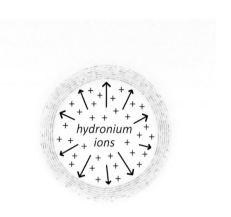

Fig. 14.2 *Droplet roundness. Hydronium ions create pressure by pushing against the EZ wall. That pressure promotes roundness.*

From Droplet to Bubble

The embryonic droplet structure just described will be central to what follows. But before we continue, let us first consider whether that vesicle is durable enough to survive. In fact, small vesicles are not particularly stable.

Absorbed energy may alter the vesicle. Suppose the vesicle absorbs radiant energy. The EZ will build. As the EZ builds, the vesicle's internal hydronium-ion concentration will build as well. More hydronium ions will raise the internal pressure. The EZ shell can sustain that increased pressure up to a point; if the pressure exceeds some critical threshold, then the EZ layers would likely begin shearing past one another, thereby expanding the vesicle (**Fig. 14.3**).

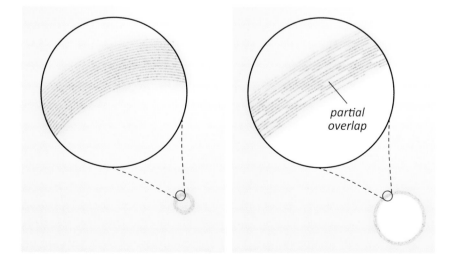

Fig. 14.3 *Vesicle expansion. To accommodate the pressure-induced expansion, EZ layers slide over one another to partially overlap.*

Vessel expansion need not be catastrophic; it can occur incrementally. The outermost layer must give way first to make room for other pressured layers to expand. Once the outer layer breaks, its segments can slip past the next layer, settling into one of the many locations where it can anchor to opposite charges. There it stabilizes. Then the same process happens to the next layer, etc. In this way, the vesicle can expand incrementally, without necessarily breaking apart (although that can happen). Substantial vesicle growth can occur in this measured way.

What happens next?

Consider the water molecules inside the vesicle. As bystanders, those water molecules sit amongst the scattered hydronium ions that perform the expansionary work. The water molecules experience pressure because the enveloping membrane presses inward upon them in the same way that a balloon presses on the gas within. If the vesicle

Bubbles in Champagne Glasses

In order to make bubble patterns visually enticing, champagne manufacturers purposefully etch patterned defects onto the insides of their glasses. The bubbles issue from those etched sites. Defects may also come from asperities situated naturally on the inside of the glass. All of those defect sites produce endless streams of bubbles, as shown in the accompanying figure.

In the case of champagne (or fizzy water), the bubbles contain not only water vapor but also CO_2. The cheery bubbles remain: Etched asperities nucleate EZs. The associated protons may then draw the negative HCO_3^- ions that form from the dissolved CO_2. The resulting CO_2-rich bubbles then rise from the asperities.

If the asperities are artfully etched, then the bubble patterns' charm may tempt you to buy more champagne.

bubble from fixed loci

238

sheath suddenly expands, then that pressure suddenly diminishes — the water molecules inside feel a reduction of pressure.

Pressure change can induce a change of phase. Pressurizing a vapor, for example, may turn that vapor into liquid; reducing the pressure may convert the liquid back into vapor. This same common principle may apply here: as the vesicle expands and diminishes the pressure on the water inside, the liquid water may convert into a vapor.

As a result of this change of phase, the droplet transforms into a bubble (**Fig. 14.4**).

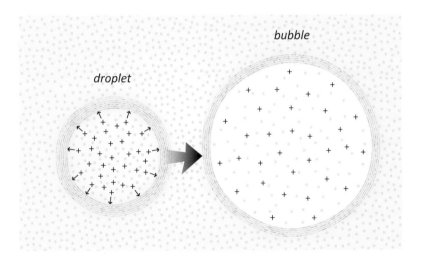

Fig. 14.4 *Droplet-to-bubble transition. Pressure expands the vesicle; water molecules experience reduced pressure, which may turn the liquid water into water vapor.*

It seems we have made some progress, at least theoretically. We began with the inevitability of EZ formation in containers of water. From that formation came droplet formation. Then came a droplet-to-bubble transition: if a droplet absorbs enough radiant energy, then internal charge elevates the pressure until the droplet expands, driving the transition to vapor and giving rise to the bubble. In this way, a bubble is born.

The question arises whether this logically inevitable progression has more than just whimsy supporting it. Does this sequence of events actually take place? Having established one of the two necessary features for the proposed sequence in the previous chapter, EZ shells, we advance to the second question: do those spherical shells really envelop positive charges?

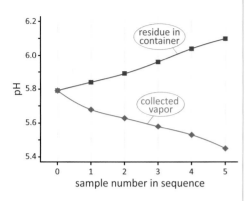

Fig. 14.5 *Results of boiling experiments. Blue curve indicates pH of the vapor, collected after various durations of time. Red curve indicates pH of water remaining in the container following boiling.*

Are Protons Really Present Inside Vesicles?

The most direct strategy for probing the insides of vesicles is simple: collect their contents to see whether the gathered molecules contain protons. We followed this strategy using the expedient of boiling. When water boils, bubbles break at the surface and their contents get expelled as vapor. We collected that vapor, condensed it into liquid, and measured the liquid's pH. The pH progressively diminished with boiling time. This meant that the insides of the bursting bubbles contained positive charge (**Fig. 14.5,** *blue curve*).

Meanwhile, as the container loses positive charge, that loss should be reflected in the residue; the residual water's pH should progressively increase. That expectation was confirmed (**Fig. 14.5,** *red curve*). Thus, protons moved from the water to the vapor, presumably as the bubbles burst and they poured from inside the bubbles.

A second test for interior protons used infrared imaging. If a bubble's interior contained water alone, without protons, then that water should not radiate much differently from the water around it. However, the interior radiates substantially more than the exterior (**Fig. 14.6**). Extra radiant energy implies active charge movement (Chapter 10) — a feature anticipated if concentrated protons jitter inside the bubble.

Fig. 14.6 *Infrared image of surface bubbles. Bubbles were created by blowing air into a surfactant solution (TWEEN 20/water). Bubble interior (orange) generates more infrared energy than the exterior (blue). Darkness of thin bubble boundaries is expected — EZs generate little infrared energy. Scale in °C.*

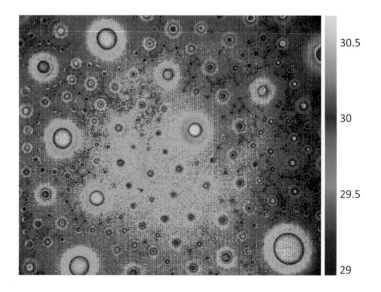

A third test for interior protons checked whether bubble breakage caused proton dispersal. If the interior radiant energy arose from some experimental artifact and not from protons, then that energy might simply vanish as the bubble breaks. On the other hand, if it came from interior hydronium ions, then breakage should disperse those ions. **Figure 14.7** confirms the dispersal. As the bubble breaks, the high-energy zone spreads, eventually disappearing as the protons mix with the water. It appears that the radiant energy comes from real material inside the vesicle.

From the evidence above, it seems clear that hydronium ions do exist inside the bubble. For the proposed mechanism, the presence of that positive charge is key: it is the essential ingredient for building pressure and driving the transition from droplet to bubble.

Vesicle Interactions: The Zipping Mechanism

Having confirmed the presence of positive charge inside an enveloping shell-like EZ, we next consider intermediate stages in the proposed bubble-formation process: how vesicles fuse with one another.

Vesicle fusion must be an integral feature of the bubble-formation process because micron-sized droplets cannot transition directly into centimeter-sized bubbles; the size gap is much too large. Phased growth is required, and that's where coalescence comes into play. The progressive combination of smaller vesicles, creating new, larger vesicles, would finally transition into the centimeter-sized bubbles seen during boiling.

To understand the principles underlying such fusion, consider first what happens in a simpler situation: when a single droplet meets a hydrophilic surface. Most hydrophilic surfaces already bear some EZ layers drawn from atmospheric humidity (Chapter 11); hence, EZ layers should line both the droplet and the hydrophilic surface. The droplet-surface interface simplifies to the interaction between a curved EZ and a straight EZ.

Suppose that droplet falls onto the hydrophilic surface. As the droplet approaches, what matters most are the positive and negative

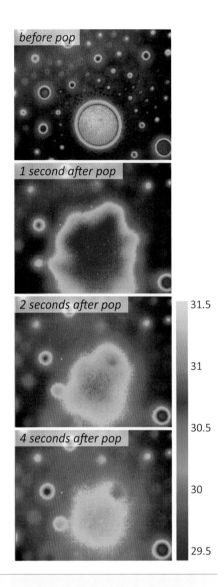

Fig. 14.7 *Bubble collapse. Similar to Figure 14.6, except that one bubble pops spontaneously. Note the dispersal of the zone of high radiant energy. Scale in °C.*

charges that make up the respective EZ layers (**Fig. 14.8**). Opposite charges will attract, slightly nudging the droplet's EZ into alignment with the surface EZ. The droplet will then stick to the surface by the attractive force of opposite charges.

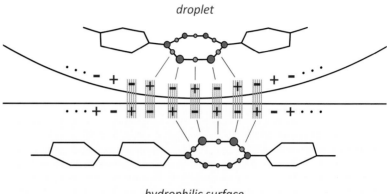

droplet

hydrophilic surface

Fig. 14.8 *Droplet EZ interacting with surface EZ. Opposite charges attract and align.*

Initially, sticking occurs only at a point, because of the droplet's curvature (**Fig. 14.8**). On the other hand, that tenuous bond is merely the starting point: charges flanking the bond point continue to attract opposite charges on the facing EZ; the respective surfaces will thereby knit together like the closing of a zipper (**Fig. 14.9**). EZ fusion will create a zone of flatness.

How much that EZ zipper will close depends on the pressure inside the droplet. Pressure promotes roundness, while zipping promotes flatness. The EZs will continue to zip together until the flattening force balances the force sustaining roundness. The result will look something like the bottom panel of **Figure 14.9**.

The zippering mechanism is essentially the reverse of the mechanism that generates vesicles. In the zippering mechanism, the spherical EZ flattens (**Fig. 14.9**). In the vesicle-generating mechanism, the flat EZ becomes spherical (**Fig. 14.1**). The two processes seem symmetrical.

Zippering provides a mechanism by which EZ shells can fuse. All vesicles have EZ shells. Therefore, the zippering mechanism can help us understand how two shells combine to produce one larger vesicle — or at least the first step in the process.

i.

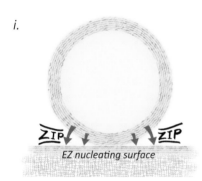

ZIP   ZIP

EZ nucleating surface

ii.

EZ nucleating surface

Fig. 14.9 *Droplet adhesion through zippering. Through local attractive forces, droplet EZ coalesces with surface EZ, forming a flat bottom.*

The Hydrophilic-Hydrophobic Paradox: How Much Do You Love Your Vesicle?

By now you probably know the terminology: Surfaces that spread water are referred to as hydrophilic, or water loving: the water clings like a lover in embrace. By contrast, surfaces that make water bead up into droplets are called hydrophobic, or water hating: they detest water. Sometimes they detest water so intensely that freshly poured water will retract by curling into fully spherical balls. A classic example of this is the lotus leaf: water dropped onto lotus leaves will recoil into spheres, which promptly roll off, leaving the leaf dry.

In order to classify one surface from another, scientists employ an artifice based on droplet shape. If the droplet remains spherical, then the surface is classified as hydrophobic; if the droplet spreads out (forming EZ layers), then the surface is classified as hydrophilic. That's simple enough. However, a confounding issue is that droplet shape most often lies somewhere in between those two extremes (see figure). The droplet will generally retain a roughly spherical shape, but with some lateral spread and flattened bottom (*center panel*).

The accepted solution to the classification problem lies in specifying the degree of hydrophilicity. Consistent geometry makes this possible: The bottom of the vesicle flattens against the surface; flattening permits easy construction of tangents, which in turn allows you to define a contact angle. Smaller contact angles denote more hydrophilicity (*left*); larger angles denote more hydrophobicity (*right*).

Why contact angle is a reasonable measure of hydrophilicity can now be appreciated. If the material is very hydrophilic with a highly charged EZ, then the zippering capacity will be high; the droplet will flatten maximally, yielding a small contact angle (*left*). If EZ charges are sparser, then the modest drawing force will produce a more limited flattening (center). And if the material is hydrophobic and no EZ forms at all, then no flattening can take place — in which case the contact angle will be large (*right panel*). So the contact-angle classification follows reasonably from this basic understanding.

This explanation of relative hydrophilicity implies something useful to remember: hydrophobicity is nothing more than the absence of hydrophilicity; i.e., hydrophobic surfaces fail to interact significantly with water to form EZs. Hydrophobicity is therefore not a characteristic in and of itself; it merely reflects the absence of another characteristic.

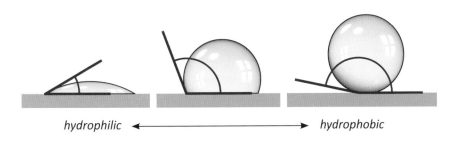

hydrophilic ⟷ hydrophobic

Droplet shape for extremely hydrophilic surface (left), *intermediate surface* (middle), *and extremely hydrophobic surface* (right). *Contact angle is often used as a measure of surface hydrophilicity.*

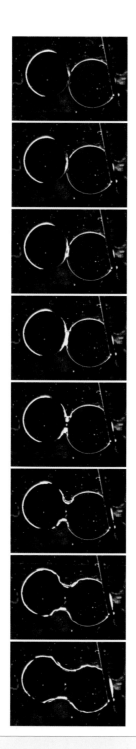

Vesicle Fusion

So far, we have dealt with a curved EZ meeting a straight EZ. More generally, both of the interacting surfaces may be curved. Two droplets may fuse into one larger droplet, or two bubbles might fuse into one larger bubble.

Such fusions presumably involve the same sort of EZ-EZ zippering just described. The zippered EZs create a plane that bisects the two vesicles. If a single larger vesicle is to emerge from that fusion, then that bisecting plane must first melt away.

To explore the melting process, we used high-speed video. We coated the inner surfaces of two glass plates with a material that produced a contact angle with water droplets of close to 90°. Holding the two glass plates parallel so that they were only slightly apart, we inserted two water droplets into the gap. The closeness of the plates, along with their coatings, forced the infused water into straight-edged discs, minimizing optical distortion. This let us track the merging of the two flattened droplets (**Fig. 14.10**).

The sequence of **Fig. 14.10** demonstrates that the two EZ membranes do indeed fuse. The coalescing EZs quickly form a straight boundary between the two vesicles, presumably by the zippering mechanism. The boundary appears to thicken substantially; however, the thickening might be illusory: any boundary-plane tilt would create the appearance of thickening because of the large optical depth of field. The main point is that the two EZ membranes coalesce along a single boundary.

Then, abruptly, that boundary fractures. The fracture commonly begins near the midpoint of the boundary plane; the material of the boundary retracts toward the periphery.

Fracturing probably results from excessive tension. To understand how this might happen, consider the equation governing tension on a thin membrane. For spherical membranes with thin walls, tension is given by the Laplace relation, $T = Pr/2$, where T is the membrane tension, P is the pressure difference across the membrane, and r is the radius. When a curved plane flattens, the radius of curvature, r, approaches infinity.

The numerator of the equation can then become extremely large. Even a minor pressure difference across the membrane, *P*, will create enormous membrane tension. That small pressure differential might arise if, say, one of the two adjoining vesicles received slightly more radiant energy than its counterpart. Any such minor pressure difference can create huge tension. Then … zap! The boundary fractures.

Videos confirm that the membrane material does retract toward the boundary plane's edges; bits of material can be seen accumulating at either edge. What becomes of this material is not immediately evident, but we can draw some inferences. Since EZs stick to one another, the retracted EZ material will likely build onto existing EZ material. Vesicle wall regions should therefore thicken and become stiffer. This stiffening may explain why some otherwise round vesicle wall segments seem to straighten following boundary breakdown (**Fig. 14.10,** *last panel*).

By the sequence of actions described above, two vesicles become one. Shell-to-shell interactions dominate the process. Interiors do matter, but mainly for pressurizing the vesicle and thereby limiting the growth of the intermediary boundary. Beyond that, the interiors don't matter much; they could just as easily be liquid or vapor, the fusion process being much the same. Hence, the same general mechanism should explain droplet fusion and bubble fusion alike.

This mechanistic universality might explain curious phenomena: droplets lodged within bubbles and bubbles lodged within droplets (**Fig. 14.11**). The common zippering process may explain the existence of these compound structures.

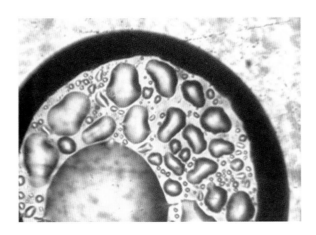

Fig. 14.11 *Compound vesicles: droplets within a bubble. The large droplet at the lower left resulted from the fusion of smaller droplets. Incompletely fused droplets may explain the observed concavities.*

To reiterate, the zippering mechanism can explain the fusion of vesicles. Once two vesicles touch, fusion is practically inevitable; the vesicles will merge into a single larger vesicle. Successive fusions, then, explain how miniscule vesicles can eventually grow into the large droplets or bubbles that we may commonly witness.

Fusion Enhances Stability and Inevitability

Fusion is important, not only because it promotes growth, but also because *it promotes durability*. The reason is purely geometric. When vesicles of similar size fuse into a single larger vesicle, shell mass approximately doubles. Shell surface area, on the other hand, increases by less than two times (try the math) — which means that some of the shells' material must go toward thickening. The new vesicle must have a thicker shell than its progenitors, which should make the new, larger vesicle more robust: it should withstand higher pressures without giving way. The larger vesicle should be more stable.

A possible glitch: more robust walls might not promote stability if the new vesicle's pressure were higher than that of its progenitors. However, that's not the case. Pressure must depend on charge concentration. When two vesicles with equal charge densities fuse, the charge density does not change: volume grows by two times, and so does the number of charges. Hence, vesicle pressure remains unchanged.

Although the new, larger vesicle experiences no change of internal pressure, its walls become thicker — which stabilizes it. By this line of reasoning, stability continues to increase with each successive fusion. This may explain why small vesicles can be short lived (and therefore difficult to spot), while larger vesicles are easy to track. The bigger ones can fend off the forces of destruction better than the smaller ones. Bigger vesicles are more robust.

This paradigm holds for droplets and bubbles alike. By merging, small droplets become more robust; so do small bubbles. All fusions promote durability, and durability, in turn, increases the likelihood of survival for additional fusions. Under appropriate conditions (see below), fusion-based growth should be inevitable — smaller vesicles always leading to larger ones.

Bringing Water to the Boil

Fusion, as you will see, is what makes boiling possible. If you watch carefully as you turn up the heat beneath a pot of water, you'll witness the succession of growth stages that eventually lead to boiling:

First you may see occasional small vesicles, which seem to mysteriously vanish; presumably, they pop inside the water. Then larger vesicles begin appearing in greater numbers. Soon they become sufficiently robust and concentrated to ensure their coalescence into larger vesicles. Eventually those larger vesicles transition into bubbles, which may break through the surface and release vapor into the air. That's how we know the water is boiling. The event is familiar, but watching it develop can lend a fresh sense of mystery — like watching witches prepare a secret brew.

• So how exactly do micrometer-sized droplets transition into centimeter-sized bubbles? The diameter ratio is about 10,000, and the volume ratio is therefore 1,000,000,000,000 — obviously too vast for a single transition. The growth must occur in stages. Here the stability feature comes into play. Since the coalesced vesicles are more robust than their progenitors, larger units more likely survive. The larger the vesicles get, the more likely it becomes that they will reach the stage of the bubble transition without popping along the way.

• Whether a droplet can eventually reach that critical point depends on ambient conditions. Since reaching the critical point depends on earlier fusions, and earlier fusions depend on vesicle concentration, then getting a lot of vesicles forming at the same time should do the trick. That feat depends on ambient energy input.

• To grow a lot of vesicles, radiant energy input must be sufficient. That's not necessarily the case early during heating, when the energy comes from the heat source alone. As heating continues, however, radiant energy comes from not only the heat source but also the warmed water. Both sources contribute. When the summed energy grows high enough, a threshold is crossed: vesicles then become numerous enough to ensure successive fusions and successful transition into bubbles. The bubbles themselves may then coalesce to form larger bubbles, and boiling is at hand (**Fig. 14.12**).

i. Vesicles meet.

ii. Vesicles fuse; EZ thickens. Multiple fusions increase volume.

iii. Vesicle expands from high energy input; vesicle transitions into bubble, with vapor inside.

iv. Bubbles meet.

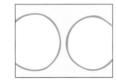

v. Bubbles fuse; multiple fusions increase volume.

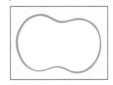

vi. Filled with vapor, large bubbles break through the surface.

Fig. 14.12 *The processes leading to boiling during high radiant energy input.*

From this analysis, it would appear that the critical variable in the equation of boiling is *not* temperature; *more fundamental is vesicle concentration*. It may just happen that vesicle concentration commonly reaches threshold at temperatures close to, although not necessarily precisely[1] at 100 °C.

Curious about the possibility of reaching very high temperatures without boiling, my student Zheng Li took a smooth, asperity-free glass beaker and filled it with laboratory-grade, distilled, deionized water. EZ nucleation sites should have been rare or absent. He then applied heat. Even when the water was heated to well above the usual boiling temperature, no boiling was evident — until he threw in some dirt. Introducing those nucleation sites brought instant boiling. He got the same result when he inserted a stirring rod — instant boiling.

It seems clear from those experiments that temperature cannot be the critical factor. Throwing dirt into hot water must have reduced the water temperature, yet it caused instant boiling. Introducing those nucleation sites evidently allowed vesicle formation, which then produced the boiling. So, adequate vesicle production seems more critical than temperature.

This conclusion reconciles a curious observation made regarding garlic soup. (If you've not tasted garlic soup, I can assure you it's unexpectedly delicious — thick, creamy, and satisfying.) From the bubbling hot soup in the pot, we ladled the contents into several rough-textured ceramic bowls. The soup immediately began to cool; however, it continued to bubble. Bubbling persisted even as the soup cooled toward comfortably palatable temperatures. Presumably the asperities on the rough ceramic bowls nucleated a sufficient number of vesicles to fuse into the larger bubbles characteristic of boiling — even at temperatures far below the supposedly standard boiling temperature.

That temperature should play only a secondary role should come as no surprise considering the discussion in Chapter 10. Temperature is ambiguously defined. Given that ambiguity, it would be surprising indeed if the critical point for boiling were uniquely constrained at some fixed value of temperature.

The Sound of Boiling Water

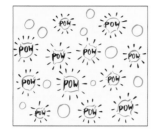 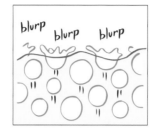 

Heat a pot of water and listen. The sound begins when you see vesicles forming at the bottom of the pot. As more vesicles appear, the volume increases, eventually reaching a loud clatter. Then, just as the water approaches the boil, the clatter mysteriously subsides into a characteristic low frequency gurgling sound — it is as though the water gremlins protesting the hell-like heat finally give up, their protests devolving into a sobbing blubber.

We hear these sounds so routinely that we ignore their presence. A good way to refresh your memory is to heat a small amount of water inside a thin-walled metal kettle. The thin walls amplify. The ordinary clatter then becomes a veritable din, which can be difficult to ignore without the help of noise-canceling headphones. The sound can be deafening.

What generates those characteristic sounds?

Sounds come from mechanical vibrations. When water is heated, vesicles begin forming. Those liquid vesicles may either fuse or fracture. Either event will generate mechanical vibrations. Fracturing will generate particularly loud pops, much like popped balloons. We interpret those pressure pops as sounds. As vesicles pop more often, the sounds will grow more frequent and therefore seem louder. Thus, hotter water will produce a more intense clatter.

Once the vesicles stop popping, the clatter will subside. That happens when vesicles begin fusing rapidly enough to transition into bubbles. The bubbles produce a quieter sound as they break through the surface and open into the air. Each bubble-breakage event contributes to that gurgling sound.

Finally comes the whistle. If boiling occurs in a whistling kettle, then the familiar high-pitched sound will eventually emit; your water is then ready for making tea. Oddly, the whistle fails to occur even when the water is very close to boiling; it's only when the water boils in earnest that the whistle begins to sound. That's awfully convenient — but why is it so?

When bubbles break, they release protons. Those protons immediately build repulsive pressure inside the kettle. That pressure pushes the vapor through the whistle at high speed (sometimes the high-speed vapor can be seen emerging from the kettle's spout), producing sound in much the same way as a clarinet produces sound. Those protons do not appear until just at the point that boiling begins. So when you hear that whistle, you can be sure that the water is genuinely boiling.

Droplets on Car Windows: My Neighbor's Radiance

Early in this chapter, I raised the question of droplets accumulating on cold glass surfaces. Orientation mattered. Now that we have some understanding of droplet behavior, I return to that issue to see whether indeed the orientation dependence might stem from my neighbor's special radiance.

A relevant issue is the nature of the moisture in the air. Although the next chapter addresses that issue, let me jump ahead by suggesting that the moisture in the air exists mainly in the form of vesicles. Those small vesicles can't be seen because they scatter very little light. However, their presence can be inferred when they condense into visible clouds.

Those airborne vesicles may also condense onto hydrophilic surfaces. Condensation occurs when the respective EZ shells cling (see **Fig. 14.8**). That's what happens on the cold automobile window, and perhaps also on the bathroom mirror as you exhale in the morning before the heat turns on. The vesicles stick. If you look carefully, you can see myriad droplets, each one clinging to the glass surface.

Driving those droplets from the surface requires radiant energy. The radiant energy builds EZs, which generate internal protons; the protons build pressure, which creates more droplet roundness. The increasing roundness diminishes the size of the zone of adhesion. Once that zone diminishes toward zero, the vesicle can no longer stick; it then returns to the atmosphere, and the window becomes dry.

The car window experience follows directly from this understanding. The window on the driver's side faces the open space, where the frigid cosmos provides no significant radiant input. So the droplets remain stuck — at least until the morning sun rises high enough to drive those droplets away. The opposite side of the car receives a continuous flow of radiant energy from my neighbor's home, which is kept warm and cozy all night. So any droplets that might like to cling to that side window are quickly driven off.

In some sense, then, my neighbor's radiance does matter — although I have been remiss in failing to transmit the good news.

Why Can You Smell the Soup?

Whether it's garlic, onion, or chicken soup, you'll catch the aroma as you approach the kitchen. Why so? With heat, it seems natural to think that the soup molecules volatilize along with the evaporating water; those molecules can then reach your nose and permit you to smell the soup.

However, something similar happens even where no heat is involved — at the seaside. Often you know you're at the seaside because you can smell the salt in the air. Salt manages to escape from the sea, presumably by evaporation. The volatilized salt can often make it all the way up to the clouds. In fact, the quantity up there is sufficient to lead scientists to speculate that the salt can actually "seed" cloud formation.

In both examples, something from a body of water makes its way to a distant place. The first example involves heating; the second does not. If a common carrier mechanism exists, then that mechanism should not be heat-induced volatilization. A common carrier might be the vesicle. Vesicles certainly form in the simmering soup, nucleated by asperities. Vesicles also exist in the breaking seas, in the form of droplets blown by wind.

Might those vesicles carry the suspect molecules?

To see how that might happen, consider the soup. As vesicles build, the EZ shells will encompass whatever liquid happens to lie in proximity. Typically, that liquid is water (with protons); however, ingredients of the day's soup may also lie near the newly forming EZ. In that case, the vesicles would contain soup molecules.

Grown from those soup-containing vesicles, the mature bubbles that pop open at the surface may then release the aromatic molecules. So you can smell the soup. Or, the molecules may simply remain inside the evaporating vesicles — which you then breathe. Even as the soup (or any food) cools, you may still smell it if the vesicles continue to evaporate (Chapter 15). The vesicles hold the smell.

Summary

This chapter dealt principally with vesicles coalescing with other vesicles. The mechanistic protagonist was zippering. Vesicle EZs zipper together, creating flat boundaries between spheres.

A flat boundary of that sort should easily give way, leading to the formation of a single larger vesicle with thickened walls. The thickened walls make the new vesicle more robust. Durability increases with each successive merger, improving the odds that the larger vesicle will survive long enough to produce even larger vesicles. This iterative process fosters vesicle growth.

At some stage, the vesicles' liquid interiors may turn to vapor. This happens if the vesicle captures enough incident radiant energy. That energy increases the number of hydronium ions inside the vesicle, which raises internal pressure. If the pressure grows sufficiently, then the shell may give way, leading to vesicle expansion. When that happens, any contained water molecules would experience an abruptly lowered pressure, prompting their conversion into vapor. Vapor-filled vesicles rising to the surface can lead to boiling.

The critical threshold for boiling may depend less on temperature than on the concentration of vesicles. When vesicles appear in high enough concentration — as they do in intensely heated water — vesicle mergers occur frequently. The merged vesicles become increasingly robust. At some point, continued growth becomes virtually inevitable. Then large bubbles may break through the surface, creating the phenomenon of boiling.

The zippering mechanism that governs vesicle coalescence also has practical consequences. When a droplet rests on a flat surface, its bottom flattens by the same zippering mechanism. The more hydrophilic the underlying surface, the more the droplet flattens and the lower its profile. Measured in terms of the so-called contact angle, the variation of droplet profile provides a convenient way to judge the degree of surface hydrophilicity.

Vesicles are critical not only for boiling but also for evaporation. The next chapter shows how, with surprising — and perhaps even dazzling — evidence.

15 Clouds from Coffee: The Remarkable Nature of Evaporation

Starbucks did not brew the first cup of coffee. According to legend, the honor belongs to a 13th century Ethiopian goat herder named Kaldi. Kaldi noticed one day that his goats seemed uncharacteristically energized; they had just nibbled on some bright red berries. Kaldi chewed on a few of those berries himself, quickly confirming their potent energizing effect.

Proud of his discovery, Kaldi brought a few berries to the local Muslim holy man. But the holy man was not impressed. He registered his disapproval by angrily casting the berries into the fire. None of that!

Soon, however, the billowing aroma from those roasting berries began to entice. Ever curious, Kaldi surreptitiously rescued some of those burned berries. He brought them home, ground them, dissolved the grounds in hot water — and voila! — the world's first cup of coffee.

Hot coffee impacts all of our senses, including sight. Vapors unfurl like cobras from a snake charmer's basket (**Fig. 15.1**). The image of **Figure 15.1** will not surprise you, but it should: by conventional thinking, the vapor ought not to be visibly detectable, as substances in gaseous form can't usually be seen.

What is required to make something visible? Visibility depends on the scattering of light. The coffee vapor is visible because the vesicles that comprise the vapor scatter incoming light; your eye detects that light. The amount of light scattered depends on vesicle size: to scatter appreciably, the vesicle's diameter must be at least the wavelength of the incident light, or roughly half a micrometer. Each of the vesicles making up the vapor must therefore comprise billions of water molecules.

There's more. If you look at **Figure 15.1**, you will notice that the vapor does not rise uniformly; a series of "puffs" ascend, one after another. The surface seems to emit vapor puffs one at a time. Each puff comprises numerous vesicles, and each vesicle comprises huge

Fig. 15.1 *Vapor rises in a series of puffs and thin streams.*

numbers of water molecules. Hence, astronomical numbers of water molecules rise with each belch.

Also emerging from the warm liquid are narrow streams of vapor (visible on the right of **Figure 15.1**). The streams look like fragile strands of spaghetti pulled from the liquid; they maintain their integrity as they rise. Being visible, those strands must comprise many light-scattering elements, perhaps strings of water-containing vesicles.

Those distinct vapor patterns rise not only from hot coffee. A student of mine noticed similar patterns at an outdoor bath in Asia. Puffs of vapor rose directly from the warm water, ascending one after another in rapid succession. In fact, such patterns are common: any hot drink will produce them. Discrete, cloud-like puffs rise in succession, with little of consequence happening in between.

Conventional wisdom asserts that liquids evaporate one molecule at a time: a random "kick" of kinetic energy separates a surface molecule from the liquid. Many such molecules may then "condense" into the visible clouds that rise into the cooler air above. Why, however, those scattered molecules should condense instantly as they pass into the air remains unclear, as does why the condensation takes the form of discrete puffs rather than one long continuous cloud.

This chapter will reconsider the process of evaporation, using what we know about the nature of water. We first explore the anatomy of these clouds rising from warm liquids to see what's inside. Then, since they emerge directly from the liquid, we ask whether such clouds reflect corresponding patterns in the water. If so, then why do the clouds rise in successive puffs rather than continuously? And what happens to the cloud's vesicles after the cloud disperses?

In short, we will investigate the fundamental nature of evaporation.

The Anatomy of Rising Vapor

To examine the vapor in the laboratory, we used laser illumination. A prism fanned the laser beam into a horizontal sheet of light, which we positioned parallel to and above a container of warmed water, as close as possible without touching. That allowed us to capture video of the rising vapor just as it emerged from the water.[1]

I recall a brash young undergraduate waltzing into my office to show me the results of that experiment. I was flabbergasted. His video frames showed that the emerging clouds (**Fig. 15.1**) were not amorphous; their horizontal cross-sections contained distinct mosaic-like features (**Fig. 15.2**). Ringed structures abutted other ringed structures to form pretzel-like mosaics. The rest was empty — vapor free. *The mosaic boundaries contained all of the evaporating water.*

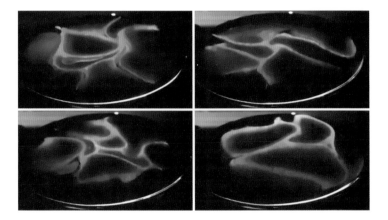

Fig. 15.2 *Examples of vapor rising from warm water. The white rings represent high concentrations of vesicles, which form the visible vapor.*

Although those mosaics seemed planar in single frames, successive frames revealed otherwise. A pretzel shape seen in one frame would persist over multiple frames with only subtle variation, usually for a span of one or two seconds. Then that pretzel would vanish. Later, an entirely different pretzel would appear, persisting again for a second or two as the vapor cloud rose past the illuminated plane.

Those observations made it clear that each pretzel must extend vertically like a stack of pretzels. Put another way, the vapor cloud had the appearance of a bundle of closely packed tubes rising vertically out of the warm liquid.

The rising vapor pattern is not rigid. Suffering impact from local convective flows, the multitubular structure inevitably distorts as it drifts upward. From a distance, that structure may look like a shapeless cloud, but voids may still be detectable as dark holes (e.g., faintly visible in lower cloud of **Figure 15.1**).

Dolphin Rings

Rings of vapor-like structures have been seen elsewhere. Dolphins exhale them. The dolphins then toy endlessly with the vaporous rings, seemingly for amusement. When the rings finally dissipate they break into myriad tiny vesicles.

Charming videos capture the fun.[w1]

Utterly astonished by these findings, we found ourselves like children experiencing the world for the first time, eagerly seeking even more surprises. We found them. We soon noticed that the tubes rose only from limited regions of the surface. One region might emit a cloud or unleash a strand, while regions immediately adjacent might produce nothing at all — no detectable evaporation. The discharging zones might shift over time, but, at any given point in time, a cloud would emerge only from limited regions of the surface.

These observations amazed us. We knew that the vapor must comprise vesicles of substantial size in order to be visible; we also felt that we understood something about the vesicles' structure (Chapter 14). However, the images implied something more. Those vesicles seemed to self-organize into large tubes, already organized as they emerged from the water. Several tubes could evidently hold together while rising into the air — although the copious vesicles that made up the tubes presumably disperse as the visible cloud dissipates well above the liquid.

Tubular vapor patterns cannot arise magically. Since the vapor patterns emerge directly from the water, you'd think that the water might contain corresponding structural patterns, which then spawn the vapor patterns. Curiosity spurred us on: might such water patterns exist?

Spatial Patterns in Liquids

You've peered at the surface of warm water many times. It looks perfectly flat and featureless. However, experimental observations tell a different story. For instance, infrared images obtained from above the warm water reveal ring-like mosaic structures much like the structures seen in the vapor. **Figure 15.3** shows an example. Later (**Fig. 15.11**), we will present evidence that those rings are the tops of tubular structures that project downward into the water.

The dark boundaries in the water image correspond to the light boundaries of the vapor image (compare **Fig. 15.2** and **Fig. 15.3**). Both boundaries contain water. If several of the contiguous dark rings visible in the water could somehow escape to the air above, they would create vapor patterns like those we observed. Of course, the water in the container beneath exists as liquid, while in the cloud above it is vapor — an issue that we must address. Nevertheless, the correspondence between liquid patterns and vapor patterns seems too obvious to ignore.

We also saw some relatively small boundary rings (e.g., *upper right*, **Fig. 15.3**); if such miniscule rings were to escape the liquid, they could create the spaghetti-like strands that also characterize the vapor (**Fig. 15.1**).

So the structural features discovered the vapor pattern can also be found in the water. This correspondence exists over a wide range of temperatures. At higher temperatures, the liquid cells are smaller, more dynamic, and more abundant (**Fig. 15.4**). Those features make sense: the higher evaporation rates at elevated temperatures imply more vesicles evaporating from the water and, therefore, a greater abundance of more dynamic vesicles inside the water. Thus, vapor and water patterns correlate.

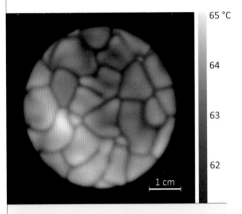

Fig. 15.3 *Infrared image of a dish of warm water, taken from above. Equivalent temperature scale at right.*

Fig. 15.4 *Infrared camera recordings of water surface patterns at a series of temperatures.*

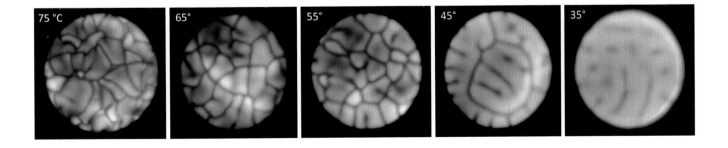

The Origin of Mosaic Patterns in Water

What might create the mosaic patterns in water?

We observed the patterns just discussed using an infrared camera. Darker parts of the image represent areas emitting less infrared energy. That means the boundaries radiate less infrared energy than the regions they enclose. One might say the boundaries are "cooler."

Indeed, the conventional interpretation of infrared images rests on temperature. That interpretation has become so standard that IR camera manufacturers provide handy temperature scales, such as the one seen on the right side of **Figure 15.3**. According to that scale, the image's boundary temperature would hover near 62 °C, while the lighter interior zones would be closer to 64 or 65 °C. The reference scale allows for a convenient interpretation — albeit one that may mislead, as I will argue in a moment.

Patterns such as those in **Figures 15.3** and **15.4**, known as Rayleigh-Bénard cells, are actually well known to aficionados. Such cells have been explored in many liquids and, to a limited extent, in water.[2]

These cells are usually thought to reflect sharp temperature gradients. The argument goes as follows: Heated water at the bottom of the container should be less dense than the water above; therefore, the bottom water should rise. When that water reaches the top and evaporates, nearby water molecules cool — evaporation is a cooling process. The cooled, denser water promptly falls. Falling takes place at the periphery of each cell, creating the cool (dark) boundary rings. Thus, the conventional interpretation provides a reasonable framework for understanding. Indeed, the up-down flows can be seen.

However, other interpretations are also possible, especially given the ambiguities attending the use of temperature (Chapter 10). One plausible alternative is based on order. A more ordered boundary material might radiate less than the region it encompasses. Charges would move less in the ordered zone, and that restriction of charge movement would produce less infrared radiation.

In support of this alternative, please recall the infrared image of **Figure 3.14**. The EZ is darker than the adjacent bulk water because its orderly structure radiates less infrared energy. The same could apply here. The dark mosaic boundary might comprise EZ material, radiating less infrared because of its liquid-crystalline stability.

To determine which option holds more promise, we adopted a simple strategy. We checked whether the mosaic patterns could be seen with ordinary visible light. Visible-light cameras create images based on optical characteristics, not thermal characteristics. (Actually, temperature does slightly affect water's optical properties, but the effect is negligible over the relevant span of several degrees.) In any event, if we could see the mosaic pattern with our naked eye or capture it with an ordinary camera, it would weaken explanations based on temperature, while making EZ-material based explanations more likely.

The images of **Figure 15.5** confirm that the patterns are perfectly detectable with visible light. Those patterns may seem less distinct than the ones obtained with an IR camera, but you can see them with your naked eye or record them with an ordinary camera.

The photograph in **Figure 15.6** shows the pattern even more clearly. We obtained this image of a pan of warm water by illuminating it with ordinary visible light at a very low angle of incidence. Contrast was obtained from the light scattered by surface vesicles, which evidently concentrate differently in the lighter regions than the darker ones. Images of mosaic patterns obtained this way are essentially identical to those obtained simultaneously with infrared imaging.[1]

Evidently, the boundaries visible in all these images comprise something distinctly different from bulk water. A good candidate is EZ water. The EZ's optical properties differ from those of bulk water in at least two ways: First, light absorption differs.[3] Second, the refractive index differs — the EZ has a refractive index about 10 percent higher than that of bulk water.[4,5] These optical differences could easily produce the visible contrast between EZ and non-EZ regions.

The EZ-centered interpretation seems to fit nicely. However, a loose end remains. Remember the vertical flows? Those flows were vital to

Fig. 15.5 *Mosaic pattern in a pot of warm water* (top) *and in a cup of warm water* (bottom). *Photos taken with ordinary cameras.*

Fig. 15.6 *Mosaic pattern seen using low-angle visible light.*

the conventional interpretation, and they certainly do exist. Do those flows fit within the new interpretive framework? If so, then what creates them and what role might they play?

EZ Material and Characteristic Flows

To gather information about the vertical flows, we looked into the mosaic patterns visible in various water-containing liquids. Soups seemed convenient because suspended food particles are often distinctly visible and therefore trackable. Warm miso soup proved especially suitable, as it shows mosaic boundary lines similar to those in the figures above (see **Figure 15.7**).

Fig. 15.7 *Warm miso soup with visible mosaic pattern.*

Observing mosaic patterns in miso soup became a preoccupation in our laboratory. Everyone began looking. Close inspection revealed two salient features. First, the boundary lines were more transparent than the areas they enclosed; those boundary lines apparently exclude soup particles, which remain suspended within the cells. This exclusionary feature lent immediate strength to the notion that the boundary comprises some kind of EZ material.

The second observable feature was that the soup particles constantly flow. They flow upward within each cell and downward near the cell's periphery. The flows resemble circulating baths. The mosaic boundaries do not visibly participate in those flows; they merely enclose each circulating area. So the classically anticipated up-down flows are indeed present; however, as far as we could tell, those flows *do not involve the boundaries*. The supposedly "cold," dense, downward-flowing boundaries do not flow.

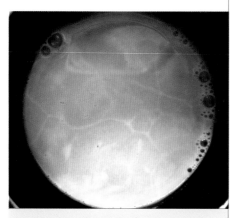

Fig. 15.8 *Mosaic pattern visible in cold mixture of milk (1% butterfat) and almond milk.*

Those boundaries are sufficiently distinct that they can be seen even without any vertical flows. **Figure 15.8** shows a 50:50 mixture of pure milk and almond milk, each component taken fresh from the refrigerator before mixing. The pattern is extremely stable. We detected no up-down flows, even in multiple attempts. At least in some liquids, then, *up-down flow is not an obligatory feature of mosaic formation*. Flow is a secondary feature, appearing most prominently at higher temperatures.

The various liquids we tested not only sustained us, but also quenched our thirst for understanding. Simple observations revealed that the mosaics were primary, the flows secondary. Further, the boundary transparency observed in the miso soup indicated the constituent material was devoid of particles; the boundaries really did exclude.

While the up-down flows seemed secondary to the mosaics, their prominence implied a role of potential significance. Particularly in warmer liquids, which have the most prominent flows and the highest evaporation, we wanted to understand how the flows fit into the overall process.

First, however, we wanted to determine the exact nature of the EZ material constituting those mosaic boundaries.

Composition of the Water Mosaic Boundaries

One option for the boundary material is the standard EZ — the kind that ordinarily lines hydrophilic surfaces. Choosing that option seems almost instinctive; however, standard EZ material does not ordinarily form large ring-like mosaic structures of the sort we had seen. So, for the moment, let us hold that option in reserve.

A second option is a mosaic built of many vesicles — similar to the vapor mosaic. Enveloped by EZ shells, vesicles could exploit the like-likes-like mechanism to self-assemble into large arrays, even networks. Those networks would exclude. The raw materials required for construction should be available, for warm water contains numerous vesicles (Chapter 14). Extremely warm water contains throngs of vesicles, and that's where mosaics are most plentiful. So the vesicle option seems promising.

In fact, individual vesicles are discernible in the mosaic. **Figure 15.9a** shows the early stages of mosaic formation in warm water; individual vesicles are visible. Vesicles are also evident in **Figure 15.9b**, which shows warm water drawn from an ordinary tap. In both instances, the ring-like mosaic boundaries seem to build from adjacent vesicles.

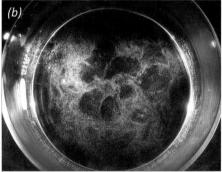

Fig. 15.9 *Incipient mosaic patterns. (a) Warm water poured into a container. Close inspection of upper right part of the image shows that the mosaic rings are formed out of individual vesicles. (b) Tap water, 60 °C, run slowly into a clear bowl. A black cloth placed beneath the bowl enhances contrast. Vesicles accumulate near boundaries, leaving the cells relatively empty.*

Contiguous vesicles also make up the vapor boundaries, as indicated earlier. Commonly referred to as aerosol droplets, those vesicles scatter light and confer visibility onto the vapor. **Figure 15.10** confirms that the vapor consists of vesicles: suitable lighting conditions make it possible to see the individual vesicles that form the vapor boundaries.

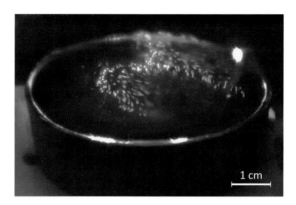

Fig. 15.10 *Vapor above warm water. Illumination is set to reveal the vapor's vesicular nature.*

Thus, liquid and vapor mosaic rings have similar structures. Both comprise contiguous vesicles arrayed into ring-like patterns. No great logical leaps are needed to deduce that the patterned arrays in the water might escape to produce the similarly patterned arrays in the vapor. Videos directly confirm just that. They show sheets of vesicles rising from the boundaries of the water mosaic to form the boundaries of the vapor mosaic.[1] *Thus, water mosaics give rise to vapor mosaics.*

Deep Mosaic Structures and Circulation

The evaporating cloud has vertical extent; it takes the form of rising tubes. If the vapor mosaic arises from the water mosaic, then the water mosaic might likewise have vertical extent, for then one structure could easily give rise to the other. In other words the water mosaics might extend downward from the surface. **Figure 15.11** confirms that expectation. The dark boundary lines of the water mosaic run downward from the surface, creating tubular mosaic structures in the water much like those of the vapor.

Fig. 15.11 *Oblique and side views of warm water obtained with an infrared camera. Downward projecting lines are evident.*

Those vertical lines are fairly dynamic. In rapidly evaporating water, they may bend or undulate toward the bottom, as though the bottoms of the lines project rather freely from the top. Something is going on continuously. Videos give some hint: they show abundant material constantly flowing downward along those vertical lines.

That downward-flowing material is almost certainly vesicular. Since the dish in this experiment contained pure water alone, some kind of water must bear responsibility for the downward-flowing material. Vesicles, abundantly present in warm water, are the only realistic candidates. It seems clear, then, that vesicles flow downward along the boundaries, creating the evident dynamism.

We can infer the nature of that vesicle flow from our miso soup observations (**Fig. 15.12**). We found that miso flowed upward within the body of the mosaic cell and downward next to the boundaries. Presumably vesicles in pure water follow a similar pattern. Vesicles that nucleate at or near the bottom of the warm water container (Chapter 14) will get caught up in the flow. The vesicles might even *drive* the flow if intense IR energy vaporizes their liquid interiors and reduces their density. Those bubble-like vesicles should rise up. In fact, the rising vesicles create visible mounds at the top center of each cell — you can see those mounds in warm miso soup. The mounds resemble a landscape of small bulging hills.

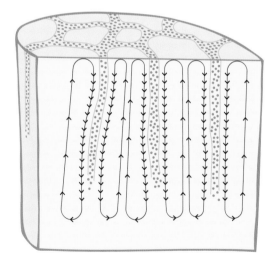

Fig. 15.12 *Circulatory flows observed in warm water. Downward flows concentrate next to the mosaic boundaries.*

Once they reach the top, the vesicles must go somewhere. Evaporating into the air is one possibility, but the vapor patterns make that option unlikely, since they show evaporation occurring from the boundaries only, and not from the cells. A reason the top vesicles don't evaporate may be that they cool from proximity to the cool air. Transitioning back to liquid-filled vesicles, they cannot take flight.

Another option for those vesicles is to return downward. Attracted to the boundaries by the like-likes-like mechanism, the vesicles would first move radially. Reaching the boundaries, they would flow downward, perhaps pushed by all the vesicles lining up behind them. That downward flow occurs next to the boundary walls — just like the particles in the miso soup. Images show those downward-flowing vesicles prominently, because they attract and concentrate near the boundaries.

The downward-flowing vesicles play a critical role: *they replenish the vesicle mosaic*. As existing mosaic material gets continually lost as vapor, it must be replaced. The descending vesicles can fulfill that role; attracted by the like-likes-like mechanism, they link to the existing wall of vesicles. With adequate replenishment, the mosaic can maintain itself, and evaporation can continue unabated. The mosaic network needs those flows to maintain its existence.

The vertical flows also make sense from an energetic perspective. The water initially becomes warm because it absorbs radiant energy. That absorbed energy drives the water far out of equilibrium with the environment. To return toward equilibrium, the water must lose energy. It can do so either by radiating energy or by doing work. The flows accomplish both. Water molecules do work as they overcome molecular friction in order to flow; this requires energy expenditure, expressed as work. The flow may also generate radiant energy as the charged vesicles move rapidly through the water. This would turn the vertical flows into a means for water to release excess energy — another example of water's role as an energy transducer (see Chapter 7).

Two messages emerge from this discussion. First, the water mosaic is a three-dimensional entity, just like the vapor mosaic; both take the form of vertical tubular composites. Second, the water mosaic tubules are renewable: as tubular complexes ascend to create the vapor, new vesicles flowing down along the mosaic wall replenish those complexes. Replenishment permits evaporative events to continue.

The Evaporative Event

What, then, triggers each evaporative puff?

This question brings to mind the larger question: what energy drives evaporation? Since water evaporates most rapidly when it receives ample heat or sunshine, a good candidate should be radiant energy. Material from the previous chapter offers a plausible scenario: radiant energy builds vesicle EZs; the EZs add interior protons, which raise internal pressure; the increased internal pressure may then expand the vesicles, turning their liquid interiors to vapor; the vapor-filled vesicles then evaporate. So radiant energy produces evaporation.

But why should those vesicles rise into the air? Vapor-filled vesicles are certainly less dense than vesicles filled with liquids; perhaps the low density causes them to rise. Interior density reduction cannot be the full story, however, for vesicles have shells. The shell consists of dense EZ material, denser even than liquid water. Depending on the shell-to-interior mass ratio, vesicles could easily remain denser than air. Something more certain than reduced density seems necessary for propelling the vesicles upward.

That upward propellant may be charge. Let me explain with a small digression.

Envision evaporated vesicles rising high into the atmosphere. Those evaporated vesicles, often called aerosol droplets, may eventually condense to form clouds. The water contained in those clouds can be massively heavy: an atmospheric science colleague estimates the weight of clouds not in terms of kilograms but in terms of easier to fathom units: elephants. In a large cumulonimbus cloud, the total aerosol droplet weight can amount to fifteen *million* elephants. That's a lot of elephants to keep suspended in the sky (and a better-than-average reason to carry a good umbrella).

Such mammoth amounts of water eventually plummet to the earth. Rain happens. Cloud vesicles seem to have two clear options — fall or don't fall. Falling must involve some reduction of the upward-directed force that ordinarily keeps the water suspended high in the sky. What if that upward force were the same as the upward force that lifts the vesicles up from the water?

This lifting force may be electrostatic, i.e., based on charges. You may recollect that vesicles carry a net negative charge (Chapter 14). Negative charge alone cannot explain lift; however the earth bears negative charge as well (Chapter 9). The earth's negative charge may repel the vesicles, pushing them upward. That upward force could help power the evaporative rise (**Fig. 15.13**).

Fig. 15.13 *Negatively charged vesicles are repelled from the surface of the earth.*

A familiar example of this electrostatic lifting force may be seen in waterfalls. Descending water creates a mist of droplets that rise upward, forming clouds. Such clouds can rise *above* the tops of the falls (**Fig. 15.14**). Since droplets cannot mechanically rebound higher than the height from which they started, some other force is implied, and a good option is electrostatic charge. The upward force arising from droplets' negative charge may be the same as the upward force that suspends the elephantine clouds — and perhaps the same as the force that propels the tubular structures upward. Those tubular structures merely need to acquire enough negative charge to facilitate the rise.

Fig. 15.14 *Niagara Falls.*[w2] *Note the ever-present cloud of water droplets rising upward.*

With this charge-based mechanism, we can appreciate why the evaporative cloud emerges in discrete puffs. The mosaic tubes bear a net negative charge, because constituent vesicles bear negative charges. Protons lying between vesicles mitigate this negativity; however, those like-like-like attractors amount to spot welds, contributing only modest amounts of positive charge. So the tubes retain net negativity. As more and more vesicles adsorb, that negativity increases and internal repulsion grows stronger. When that internal repulsive force exceeds a critical threshold, the tubule literally tears itself apart at the weakest point. The top section may then rise upward — repelled by the negatively charged vesicles below and also, ultimately, from the negative earth.

That charge-based process creates the single cloud puff. The process is *catastrophic*; i.e., it occurs because of a charge-based instability that triggers some portion of the mosaic matrix to break free. The process may then repeat, yielding the familiar succession of cloud puffs — similar to those puffs you see wafting upward from your cup of hot coffee.

Completing the Evaporative Cycle

The vapor clouds rising from hot coffee ultimately disappear; they vanish into the air above. We might rightly ask why and how those clouds vanish. Also, what happens next?

Two possible explanations for disappearance come to mind: the vesicle aggregate might break up; or the vesicles themselves might break up. Vesicle breakup is conceivable, but a physical reason seems less than obvious. Aggregate breakup is simpler to envision because the positive charges linking the vesicles could easily disperse into the air. The vesicles would then be freed — but intact.

Once freed, those vesicles should disperse. Dispersed vesicles easily escape detection: while single vesicles do scatter light, you might not notice the modest amount of light scattered by vesicles widely dispersed. In the humid air of summer, however, the picture differs. The vesicles' higher concentrations should appreciably scatter light, explaining that season's oft-seen haze. That scattered light compromises distance vision. It's like trying to see through a thin cloud.

Which suggests another point: those dispersed vesicles stand ready to form clouds. The vesicles merely need to aggregate (as they do in

Vesicle Charge and the Kelvin Water-Dropper

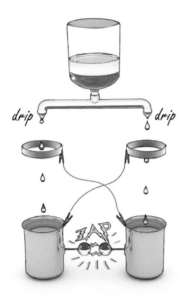

Zap! An electrical discharge between two metal containers of water, both filled from same water source. Weird — but that's the Kelvin water dropper demonstration (Chapter 1). The observable discharge bespeaks the falling droplets' ability to carry charge.

Here's how that discharge works. Suppose the first falling droplet happens to contain just a trace of net charge, let's say negative. If that droplet falls into the left bucket, then the left bucket will gain a trace of negative charge.

Now in the Kelvin experiment, all buckets and rings are metallic; they conduct. If the left container acquires a negative charge, then the right ring will also be negatively charged (see diagram). The right ring's negative charge — here's the key — will induce equal and opposite charge on the incipient water droplet hanging just above. (Induction of opposite charge comes from basic electrostatics.) The tip of that drop of water, about to fall, will therefore become positively charged. When the drop falls into the right bucket, that right bucket will gain a trace of positive charge.

The right bucket's positive charge will in turn confer positivity onto the left ring, which induces negativity in the water column just above. So the next droplet to fall on the left will bear a negative charge.

Thus, every drop falling into the left bucket will contribute negative charge while every droplet falling into the right bucket will contribute positive charge. As charge builds in each bucket, and hence each ring, the induction effect will grow stronger. Eventually, the buckets become so highly charged that they discharge via an arc between the buckets.

Apart from the impressive zap, the droplets' dynamic behavior offers a sideshow. Droplets sense the increasing charge and begin deflecting away from the bucket. The droplets may even rise upward, often missing the target bucket altogether (see figure below). This makes it clear that charge effects can be strong enough to defy gravity.

This latter observation lends support to the notion that the rise of vesicles from warm water may be electrostatically driven. Electrical forces can evidently propel droplets upward.

Droplets falling into the bucket of the Kelvin apparatus, illuminated with red light. When bucket water acquires sufficient charge, falling droplets are repelled upward.

warm water). This process need not be complicated — it requires nothing more than some positive charge — but, in the interest of brevity, I will leave a detailed discussion of this process for another time.

We have seen that vesicles provide a vehicle for continuity. The vesicles initially form in water; they rise as vapor, disperse, form clouds, and ultimately fuse with one another to create rain droplets, which return back to the earth to complete the cycle. Thus, vesicle dynamics may be a central feature of the water cycle and, hence, central to all weather.

Bug Screens and Air Flow

Despite the temptation to end this chapter here, I feel drawn to consider at least briefly what happens to vesicles dispersed in the air. Some of those vesicles might eventually condense into clouds. All other vesicles, you might guess, should remain suspended in the air, each one floating aimlessly and independently like children's blown bubbles.

As charming as this picture may seem, the free-floating scenario seems unrealistic. For at least two reasons, some kind of interaction between vesicles and air seems probable. First, the unitary vesicle is charged; charged vesicles will inevitably gravitate toward and stick to any opposite charges, and the atmosphere contains plenty of opposite charges. Second, air molecules themselves show signs of linkage — which could easily involve the charged vesicles. It is this latter possibility that I want to consider in the coming paragraphs, because that linkage is unanticipated.

To confirm these linkages, try the following experiment (**Fig. 15.15**). On a humid day, notice the breeze wafting pleasantly into an open window of your home. Now slip a bug screen into the window frame, and note the decrease of wind speed. Several colleagues told me of this phenomenon. In my own experience at home, the speed on the leeward side dropped substantially. A crude sensor showed that the speed approximately halved. You expect some drop in speed because the screening material partially blocks the air passage; but the physical material in my bug screen covers only 10 to 15 percent of the area, not at all proportional to the fall in speed. Something more seemed to be going on.

Fig. 15.15 *Ordinary window screens reduce moist air flow more than anticipated.*

Let me put the scenario into a more quantitative perspective. Air molecules are measured in nanometers, while the screen openings are measured in millimeters. That's a linear ratio of a *million* times.

To appreciate the difference of scale, imagine a bug screen with openings each the size of a mountain. Suppose you cut out one of those huge rectangular units and stand it vertically on edge (**Fig. 15.16**). Imagine that, when you now drive golf balls through the opening, you notice that the mere presence of that screen boundary is sufficient to slow down all the balls passing through — and, when you remove the boundary, the velocity returns to normal. As weird as that may seem, it's analogous to what happens as the air molecules pass through the bug screen. The size ratio is similar.

Fig. 15.16 *Bug-screen analogy. Even though the boundary is massively large, its presence slows down passing golf balls.*

Turbulence and eddies may play some role in retarding the flow, but the speed diminution seems too impressive to explain away by so localized an effect. Something else appears to be going on. An unconventional possibility: suppose the air molecules were linked to one another, forming a loose net. Then, any molecule(s) hitting the screen material would slow down all the rest of the molecules.

Of course, air molecules should be unlinked in theory: molecular independence is the very definition of a gas — an ideal gas, at least. However, the screen result demands an explanation, and the possibility that theoretically independent molecules might be linked seems worth exploring. Could some evaporated entity create linkages?

Linkages in the Air?

Evaporative entities include both the vesicles and the proton glue that rises along with them. If the vesicles disperse, then the protons might disperse as well. At first, those protons looked like attractive candidates for linkage: their positive charge could link electronegative sites on the air's nitrogen and oxygen molecules. Despite their initial promise, the proton linkages can only create molecular pairs. Pairs won't suffice; the bug screen results imply extensive air linkages.

Dwelling on those possible proton linkages nevertheless brought to mind a long-known paradox that suddenly seemed freshly relevant. That paradox is the constant nitrogen-to-oxygen ratio. By volume, dry air contains 78.09 percent nitrogen and 20.95 percent oxygen. The ratio is 3.727. While concentrations of trace gases like argon and carbon dioxide can vary widely from place to place and time to time, the ratio of oxygen to nitrogen remains stubbornly constant — to four significant figures.[w3] That's an awfully consistent ratio.

This constancy seems to hold everywhere — in cities, on farms, atop mountains, in deserts, over oceans. Even in locales like wintry Siberia, where photosynthesis surely contributes far less oxygen than in the jungles of the Amazon, still the ratio holds fast. In fact, the ratio remains so invariant that atmospheric scientists work assiduously to develop instruments that push measurement precision from four significant digits to five; succeeding might facilitate the search for even trivial oxygen differences.

A possible explanation for this consistency is that these gases' earthly turnover is trivial relative to overall atmospheric content; i.e., the vast oxygen-generating flora covering the earth might not really matter. While possible, this is not easy to reconcile with current understanding — that atmospheric oxygen actually *originated* from the flora.

Another possible explanation — and here I go far out on the limb of speculation — is that nitrogen and oxygen form stoichiometric complexes — i.e., complexes containing fixed ratios of oxygen to nitrogen. Complexes of that sort are known as gas clathrates. Gas clathrates typically contain fixed numbers of gas molecules trapped within cages of water. In the present case, the complexes would contain fixed numbers of nitrogen and oxygen molecules, electronegative entities held together by positive protons.

How many molecules?

Gas clathrates commonly contain up to dozens of molecules. In air, the nitrogen to oxygen ratio is near 4:1 by volume; if the molecule ratio were exactly 4:1, then the clathrate might contain only five molecules (**Fig. 15.17**). That's one option. Other integer ratios would yield larger numbers with different arrangements, but the essence would remain the same: stoichiometric complexes of nitrogen and oxygen.

Fig. 15.17 *Simplified schematic of possible gas linkages (not to scale). Protons might link electronegative sites on oxygen and nitrogen, creating stoichiometric linkages such as the one shown. The actual number of molecules would be larger.*

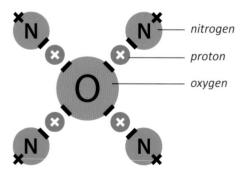

Further, if the probability of forming such a complex were high, then virtually all of the air's nitrogen and oxygen molecules could be complexed in this way; the nitrogen-to-oxygen ratio would then remain rather invariant over space and time, as observed.

The clathrate hypothesis has the advantage of defining a role for the atmosphere's known positive charge. Scientists are aware of this positive charge, but don't know its origin. The charges could originate in the protons freed during evaporation. Those freed charges could create the molecular linkages, helping to account for the gas ratio constancy.

While solving that problem, those protons still don't help with a problem we need to solve: extensive linkages among gas molecules. Here the other evaporative entity might come into play: the vesicles. Negatively charged vesicles always seek out positivity. The most abundant positive charge source is the exposed end of the nitrogen molecule on the clathrates' outer edges (see **Fig. 15.17**). Negative vesicles latching onto those positive sites could create extensive linkages (**Fig. 15.18**).

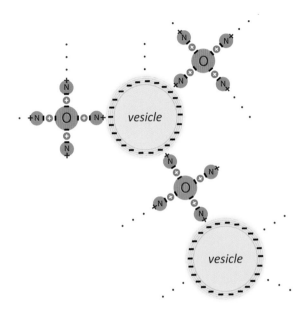

Fig. 15.18 *Vesicles might loosely link oxygen-nitrogen complexes to form a continuous structure.*

While the speculated linkages need followup by serious experimental exploration, they do have the capacity to explain the bug-screen results. Such linkages should be weak enough to escape casual detection, while having enough strength to explain why humid air is commonly described as "thick" or "heavy." The many vesicle linkages would also help explain why screens slow the flow of air most in high humidity.

Atmospheric Conductivity and Friction

Beyond solving the bug-screen problem, the proposed linkages might help explain two paradoxical but seemingly unrelated phenomena. The first is the atmosphere's remarkable ability to transmit radio signals.

As a child, I wondered how radio waves generated in Australia could reach all the way to Brooklyn. Those distant signals could be picked up by my radio and my friend's radio next door. Somehow, energy radiated from the other side of the world filled the local atmosphere. Even if those waves bounced multiple times off the ionosphere and the earth, still I could not understand how they could maintain strength over such immense distances.

My primitive crystal radio performed almost as impressively; it too could pick up radio signals sent from huge distances. Yet, *it had no battery*. Those far-traveling signals must have contained the necessary energy — energy enough even to power my headphones. Incredible! To my knowledge, a satisfying explanation has yet to emerge for this astounding feat.

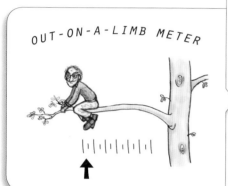

OUT-ON-A-LIMB METER

Whether the air-molecule linkages posited above might help solve the transmission problem remains speculative — even more so than the linkages' existence. However, such linkages could provide electrical continuity. A signal sent from Australia might travel along those atmospheric "wires" in much the same way that signals travel along copper wires. The signals could travel virtually everywhere. Losses would be anticipated, but since incident radiant energy continuously feeds the vesicles, those vesicles could serve as signal amplification nodes, boosting the signals everywhere just like transistors boost signals. In that case, plenty of signal could be available to power even passive crystal receivers.

Continuing in a speculative vein, I'd go a step further, to suggest that those linkages might resolve another unrelated atmospheric conundrum: why the atmosphere moves in lockstep with the earth. Consider this: The earth spins relentlessly around its own axis. Relative to the reference frame of the universe, you are whizzing around at 1,500 km per hour, twice the speed of a jet plane. The air around you evidently moves in lockstep — for if it did not, then you'd always feel a howling wind (**Fig. 15.19**, *opposite page*).

To explain why the air moves in lockstep with the earth, you might argue that, well… that's just the way it is. When the earth formed, the air spun right along with it; and, because of the air's momentum, its speed might persist undiminished — just like that of the earth. However, that explanation doesn't suffice: air velocities can change practically by the minute; therefore, other factors must outweigh any supposed inertial continuity.

Fig. 15.19 *Earth-atmosphere coupling. The atmosphere moves together with the earth's surface (top). In the absence of coupling between air molecules, the atmosphere might not move much at all (bottom). An earthbound observer would then detect an east-to-west wind of high velocity all the time.*

An alternative explanation for the lockstep rotation is mechanical coupling: atmospheric components remain weakly linked, and the linked atmospheric entity couples to the earth by friction. Rolling hills, tall buildings, and protruding mountains might drag the near-surface air along with the earth. Were the upper air molecules unlinked to those molecules beneath, they'd remain part of the cosmos. To an earth observer, those atmospheric molecules would whiz at supersonic speeds in the direction opposite the earth's rotation. That never happens.

Air and earth must therefore be mechanically coupled, even high into the atmosphere. The two entities must spin as a unit. Coupling seems difficult to explain unless air molecules are at least loosely linked to one another. Then, as lower air molecules move with the earth, so will the upper ones. Coupling is fortunate; otherwise, we might suffer the hellish vision of eternally howling super-hurricanes. Imagine a commuter flight from Chicago to New York having to buck those eternal winds!

Coupling of air molecules may also help explain why air has as much friction as it does. Think of meteors burning as they pass into the atmosphere; think of planes consuming as much fuel as they do; and think of objects falling from a tall building and reaching a terminal velocity. All of these phenomena result from air friction, which in turn results from connectedness.

Naturally, there's more to air-earth coupling than linkages and friction. I hesitate to digress even further from the chapter's topic of evaporation, but I simply cannot leave without mentioning the obvious: the earth is negative and the atmosphere is positive. They attract. Whether this attractive force is substantial enough to couple the air to the earth is a question left for future investigation; it could be a dominant factor — possibly even explaining the so-called air pressure.

The obviously speculative material of the last several sections was included mainly to raise questions rather than to answer them. The chapter's principal message is detailing the little-known yet astonishing sequence of events surrounding the evaporative process. Hopefully those evaporative events are better understood — particularly the vesicle clusters that ascend regularly from the water disguised as puffs of vapor.

Summary

Vesicles self assemble in water. They do so by means of the like-likes-like mechanism, forming extensively networked structures. These structures resemble mosaics when viewed from above; however, the mosaics are actually tubes, extending deep into the water. With sufficient absorption of radiant energy, the tubes may acquire enough negative charge to escape the water individually or collectively. The rising structures, seen as puffs of vapor, emerge one after another from the surface. Those emerging puffs are the essential elements of evaporation.

How do those evaporative events relate to the vesicle-generating processes considered in the previous chapter? Energy input is crucial. With infrared (heat) input, vesicles form in abundance; the vesicles flow vertically and thereby augment the tubular mosaics — which may then rise as vapor. More energy input means that vesicles form more quickly and mosaics rise more rapidly. Hence, higher energy input yields faster evaporation.

With even higher infrared input, vesicle production may become so rapid that vesicles have little opportunity to join the mosaics. Those vesicles may simply coalesce, transition into bubbles, and rise to the surface in the phenomenon we know as boiling. Boiling is the evaporative extreme; it is sufficiently chaotic that mosaic regularity practically disappears.

At the other end of the heating spectrum lies unheated water. There, evaporative processes can be presumed to occur as described in this chapter, but at a reduced rate. On the other hand, the room temperature condition implies a kind of stability. Accompanying this stability come several unanticipated features that we will explore in the next chapter.

16 Water Trampolines: Layering at the Water's Surface

Skipping rocks across the water's surface was once a teenage ritual. When I was a teenager, a subtle competition prevailed among rock skippers — a test of manhood whose outcome might determine your likely success in luring girls. Those who could bounce stones farthest surely qualified as the alpha males.

Why should those rocks bounce? Certainly rocks can rebound off trampolines, but the water surface would seem quite unlike any such elastic sheet. Water is a viscous liquid — rocks shouldn't easily ricochet. On the other hand, when water meets air, special features show up: EZ mosaics cover the surface and project appreciably downward into the water (see **Fig. 15.11**). So surface water differs from the bulk water beneath, raising the question of whether those surface features might sufficiently account for the skipping of rocks.

This chapter takes a deep look at the surface of water. It reveals some surprising mechanical features, which help clarify phenomena ranging from walking on water to why ships float — the latter taking us a bit deeper than Archimedes.

Surface Water Differs from Bulk Water

Finnish ski jumpers like to train year round. In winter, ample snow accommodates their needs, but summers present a challenge. The resourceful skiers nevertheless improvise: they ski on plastic tracks strategically situated so that the skiers can land in water.

The water, however, is not as accommodating as one might expect. Landing ski jumpers are prone to bone fracture unless the surface has been presoftened through vigorous bubbling. The bubbles continually break through the surface, diminishing the surface tension and permitting the skiers to land safely. Novice high divers often employ a similar strategy; few of them break their bones.

Fig. 16.1 *Evidence of water's high surface tension.*

The stiffness of the water surface should come as no surprise, for high surface tension is one of water's known anomalies. The tension is high enough to float dense objects, ranging from steel pins and paper clips to old Hungarian coins (**Fig. 16.1**).

Scientists have generally attributed water's high surface tension to extra hydrogen bonding. Surface water molecules have no binding partners above; those unmade bonds can be directed to neighbors. The extra lateral linkages increase the stiffness, yielding the high surface tension.

The surface layer containing those extra bonds would be less than one *nano*meter thick. To realize what one nanometer means in familiar terms, think of a one-millimeter slice of salami. Slice it thinner — by a thousand times; then take one of those slices and slice it in the same direction another thousand times (if you can). A flimsy film one-tenth the thickness of a cell membrane supposedly makes the difference between breaking or not breaking one's neck.

Something more significant than a few extra bonds in a thin film seems necessary to account for water's unusual surface properties.

Walking on Water

Water can support live creatures, ranging from water striders to Central American lizards. Stunning videos show Costa Rican lizards scampering across pond surfaces.[w1] Because they can walk on water, those creatures are called Jesus Christ lizards. This phenomenon makes it clear that natural water surfaces may be considerably stiffer than one might suppose.

EZ-Like Zones at the Air-Water Interface

Something more significant is indeed present at the surface: a mosaic (Chapter 15). This structure penetrates downward from the surface, creating a thick net-like layer. That net-like layer cannot help but materially affect the surface's mechanical features.

We stumbled upon this distinctive surface feature fortuitously — before we dreamt of using an infrared camera to explore its properties. In chambers containing water and microspheres, we noticed a microsphere-free zone that lined the water's upper surface.

We first saw that clear zone in beakers. The aqueous suspension inside the beaker was initially cloudy with microspheres. Soon, however, a microsphere-free zone developed just beneath the surface, remaining in evidence for an extended period of time. What impressed visitors more than that plate-like zone was the clear cylinder that emerged much later, running vertically near the middle of the beaker (see **Fig. 9.12**). That prominent vertical cylinder formed out of the more subtle disc-shaped clear zone at the top.[1]

The upper clear zone showed up again in another type of chamber, created by placing two parallel glass slides close together and sealing them around three edges so they would hold water. It resembled a narrow fish tank (**Fig. 16.2**). Under some experimental conditions stable clear zones could be seen. Suspensions of microspheres looked uniformly cloudy at first. But, within minutes, a microsphere-free zone would develop at the top. That clear zone persisted for about a day, after which all microspheres sedimented to the bottom.

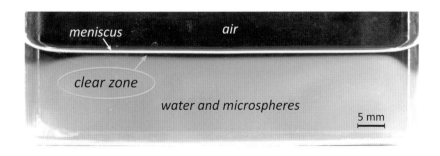

Fig. 16.2 *Clear zone near the top of a microsphere suspension in a chamber built of two parallel glass slides, sealed at left, right, and bottom.*

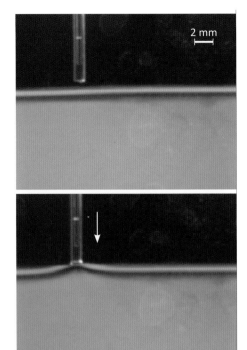

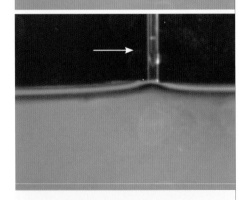

Fig. 16.3 *Glass probe lowered to touch the top surface of water. Thickness of EZ is hardly altered by that mechanical perturbation or by the probe's sideways movement.*

Thus, we saw surface clear zones in both cylindrical and rectangular chambers well before we understood that they might correspond to the mosaic-like structures. The clear zones struck us immediately as EZs of some kind. They excluded microspheres, and the tops of those zones bore the EZ's characteristic negative potential, as we later found. If the subsurface clear zone comprised EZ material, then the zone's stiffness might be sufficient to support steel pins and Hungarian coins.

We quickly confirmed the zone's high stiffness (**Fig. 16.3**). We slowly lowered a vertical glass probe toward the water surface. At some point before we expected it to touch the water, the surface lunged upward (perhaps because of induced charge) to meet the probe. During this mechanical perturbation, the clear zone just beneath the probe hardly changed thickness; nor did its thickness change as the probe was subsequently shifted from side to side. The clear zone behaved like a cohesive band — a rubber dam — stretching across the water surface. Being millions of molecular layers thick, that band should experience little difficulty supporting fairly weighty objects.

This surface band seemed to correspond to the structure seen in the previous chapter: a clear zone when viewed from the side, a mosaic structure when viewed from the top. Neither view gives the full picture; together, they reveal more comprehensively what exists immediately beneath the water's surface (**Fig. 16.4,** *opposite page*).

The previous chapter's observations were obtained mainly from studying warm water, while we observed the above-described clear zones at room temperature. If the two observations reflect the same structure, then we can surmise that the room-temperature surface features are at least qualitatively similar to those seen at elevated temperatures: mosaic-like EZs that cover the surface and extend appreciably down into the water.

The EZ mosaics comprise mainly clustered vesicles. However, they may also include standard EZ material. The standard EZ material might come from two sources. First, the mosaic's outer boundary lies just inside a container wall; the wall could nucleate a standard EZ structure, which could participate in the formation of such mosaics. Second, some vesicles could transform themselves into standard EZ material by the zippering mechanism (Chapter 14). The resulting mosaic would then seamlessly mix standard and vesicle EZs, whose relative content would depend on ambient conditions.

Ambient conditions may also determine what fraction of the surface is covered by the EZ mosaic. **Figure 16.4** shows a rather open structure with modest fractional coverage. In theory, the mosaic's openings could largely fill in. Fractional surface coverage depends on the number of vesicles, which is a balance between vesicle production, vesicle absorption onto the existing matrix, and vesicle loss through evaporation. At room temperature, the limited evaporation rate could tilt that balance toward filling the surface with EZ-containing vesicles.

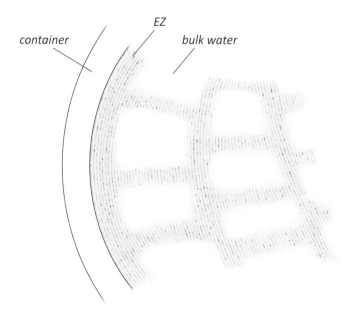

Fig. 16.4 *Looking down at the mosaic surface structure. When viewed from the side, this structure may appear as a clear zone at the top, especially if bulk water openings are relatively small.*

Quantitative uncertainties notwithstanding, the net-like mosaic should stiffen the surface. That stiffening may explain water's anomalously high surface tension.

The stiffening might also explain the resistance encountered by high divers. On the other hand, stiffness is not the full story: when skiers or divers hit the surface, the underlying water needs to get out of the way; the water must accelerate against the inertial forces that act to keep it in place. The net-like layer constrains those water molecules; it keeps them from accelerating. For divers, then, the surface mosaic presents a double obstacle: it stiffens the surface, and it constrains the water from easily getting out of the way. Happily, this surface obstacle can be eliminated by continually releasing bubbles from beneath the surface, a process common in indoor diving.

Thicker Surface Zones in Open Waters?

The observations described in the previous section were mostly made in laboratory settings. Conditions may differ in deep bodies of natural water continually exposed to relatively high levels of radiant energy. There, the surface mosaics' fractional coverage and vertical extent may differ from those in laboratory beakers. Indeed, those surface structures might extend substantially deeper.

One hint of this increased vertical extent comes from the reports of competitive freedivers. Holding their breath for up to eight or nine minutes (!), these athletes can descend to depths in excess of 100 meters before resurfacing. They consistently report a physical transition at a depth of 15 to 20 meters. Above that depth, the body seems almost neutrally buoyant, whereas below it, the body is said to sink like a stone.

That situation seems analogous to the pin in the glass of water: it may float when placed gently on the surface, but when forcibly submerged to a point beneath, the pin sinks easily. This transition point lies millimeters from the top; in the case of freedivers, the transition point seems to lie some meters beneath the surface.

A second hint of increased mosaic depth comes from naval engineers working with sonar. Sound directed downward ordinarily penetrates to the bottom of the sea. But, if the sound is directed obliquely, then it bounces off a discontinuity somewhere beneath the surface and never makes it to the bottom. The same happens from beneath: sound directed obliquely upward may never reach the top. The responsible discontinuity seems to occur at varying depths. In the shallower waters near coastlines, that depth is similar to what the freedivers report: some meters. In deeper ocean waters, the interface reportedly occurs at depths of several hundred meters or more. The source of the discontinuity remains unsettled but could correspond to the mosaic's lower boundary.

A third relevant observation comes from a shipboard study.[2] Measurements carried out in the Baltic Sea have once again revealed a vertical discontinuity. From the surface down to about 60 meters, the investigators found a practically constant oxygen level, which then, within 10 vertical meters, dropped sharply to a much lower level (**Fig. 16.5**). The high oxygen content near the surface supports the presence of EZ material, for EZs are densely packed with oxygen (Chapter 4). Moreover, the upper zone contained a salt concentration less than half the value seen farther down. Since EZs exclude salt, the low salt concentration in the upper zone is also consistent with an EZ presence.

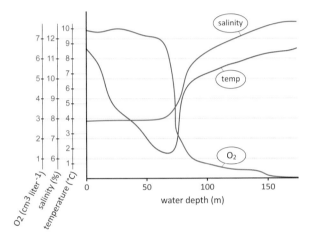

Fig. 16.5 *Vertical profiles in the ocean.*[2] *Temperature, dissolved oxygen, and salinity profiles, measured at Baltic Sea station, 26 May, 1979.*

A particularly intriguing finding of that study concerned the distribution of amino acids. As the sun bore down during the day, the concentration of dissolved amino acids progressively diminished in the top zone while increasing at lower depths; evidently, amino acids moved downward. If sunlight increases surface structuring, then the downward shift of excluded material is anticipated. That shift should reverse as the sun begins to set, and that was observed. Apparently, the quantity of excluded material waxed and waned with the sun, as expected if the responsible agent is EZ material.

This collection of evidence leads us to speculate that extensive regions of the sea's upper zone may be EZ-like. In the laboratory, such EZ-like zones extend down by millimeters, occasionally centimeters; in the sea, they may extend downward by tens of meters near shorelines, and, farther out to sea, perhaps hundreds of meters. This depth is not surprising given the abundance of radiant energy and oxygen, as well as the countless millennia available for attaining a steady state.

Although they may extend to impressive depths, those EZ-patterned zones are probably discontinuous, even beyond having small fenestrations like those shown in **Figure 16.4**. The sea is continually shifted by tides and lashed by winds; hence, EZ structures are likely to suffer multiple fractures. Moreover, the upper zone is populated by marine gels and penetrated by all manner of creatures. Hence, that upper zone may exist as a structural patchwork rather than a continuous network. Nevertheless, the thick, net-like structure should inevitably stiffen the water's surface.

Tsunamis

A thick layer of EZ material lining the ocean's surface would make it easier to explain certain phenomena — especially those involving waves.

Waves pervade the ocean surface. The current interpretive framework for waves builds on the presumption that oceans consist of bulk water alone; mass and viscosity therefore dominate. Explanations rest on arcane phenomena such as Stokes drifts, frequency dispersions, Boussinesq equations, etc.; those explanations require different approaches for different depths. The resulting wave models are intractably complex. Achieving even modest intuitive understanding requires simplifications.

In elastic media, by contrast, waves are natural: think of plucking a guitar string. Waves don't quickly damp out, as they do in

viscosity-dominated frameworks. If the surface zone of water could be modeled as an extensive elastic sheet, then propagating wave phenomena might be explainable even on an intuitive level.

The elastic sheet model would seem to fit the phenomenon: the mosaic net exists; and when modestly strained and released, the net would merely need to return quickly to its unstrained configuration. The mosaic net should satisfy that requirement.

In this interpretive vein, it is worthwhile to consider the extreme example: tsunami waves. Those towering and oft-devastating waves may circumnavigate the earth several times before finally dissipating. Envisioning such sustained propagation within the framework of a viscous liquid is not easy; friction should quickly damp out the waves. On the other hand, propagation is easier to fathom in the context of a stiff elastic sheet: perturbations can propagate swiftly over long distances. The sheet model might explain why tsunamis propagate as far as they do.

The continuous sheet model may also resolve a mystery: why the sea recedes from land just prior to the tsunami's arrival onshore. The wave crest constitutes an upward deflection (**Fig. 16.6**); upward deflection of a continuous sheet will draw that sheet inward from the periphery. In the case of the tsunami wave, the upswell will pull the edge of the sheet seaward, explaining the sea's pre-inundation withdrawal. Do keep a weather eye out for any such withdrawal!

Fig. 16.6 *Tsunami dynamics with an elastic surface sheet. An upswell will draw the elastic sheet seaward, as observed.*

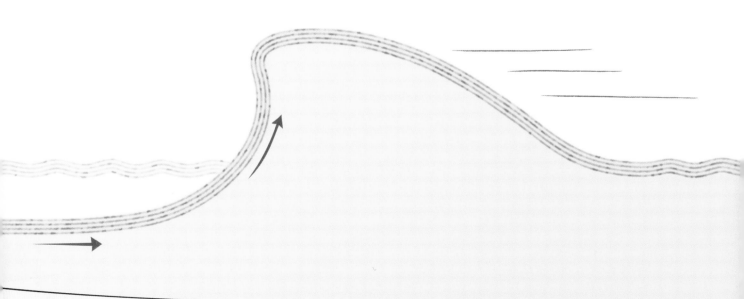

An additional expectation: the sizzle. As the upper sheet pulls seaward, some of the bulk water lying beneath the sheet may remain in place. That bed of water should be rich with protons — as would any water situated next to an EZ sheet. Now freed, those hydronium ions repelling one another should spray upward into the air like a freshly opened can of just-shaken pop. This spray may explain the oft-reported sizzle.

Submarine Detection

EZ material lying beneath the ocean's surface would create an appreciable subsurface layer. That layer should be deformable — as demonstrated in Figure 16.3. Thus, mechanical perturbations originating beneath the plate could deflect the plate, and those deflections might be detectable from above. Indeed, infrared lasers have detected water-surface bulges, known as Bernoulli humps, arising from submarines passing beneath.[w2]

Water Surface Fragility

While the elastic sheet model may be a useful expedient for understanding waves, it falls short of accurately describing the water surface. For the most part, the surface comprises packed vesicles. This packed vesicle material may suffice to explain surface elasticity, but what about surface fragility? The surface must be fragile, otherwise fish could not pass through.

Fragility can be understood in terms of thixotropy. One of those tongue-twisting words you may have encountered but perhaps can't define — or even pronounce — thixotropy refers to a particular material character. Thixotropic materials, when nudged gently, will return elastically to their initial configuration; past a certain threshold, however, they begin to flow. Think of egg white.

Egg white models the type of fragility I mean. Egg white is replete with ordered water[3] — understood now to be EZ water. Egg white's EZ water excludes, as you might expect. To see this for yourself, expose egg white to various food colorings: provided the gooey albumin is left mechanically undisturbed, the egg white will exclude those dyes.[4]

EZ material such as egg white behaves thixotropically because of the electrostatic nature of its bonds. Think of the vesicle array. Opposite charges hold the vesicles together. Those bonds should sustain small deformations without giving way; they behave more or less elastically. However, if you tug on the array enough to break those bonds, then it will fracture and vesicle material will flow. Such fracture may explain the surface material's fragility — why swimmers and fish can pass through with little difficulty.

Thixotropy may also explain why coins float when eased gently onto the water surface but sink if placed carelessly. Careless shear may fracture the surface structure, thereby allowing easy penetration; the object simply drops through. Scrupulous placement can avert this kind of surface disruption, permitting the object to remain floating on the surface.

The same principle holds for ships. Moving ships induce major shear. That shear easily breaks up the surface structure, allowing the ship to pass through without difficulty. The shear is less pronounced beneath and beside the ship, where the EZ structure remains intact but strained, not yet reaching the breakup threshold.

Ships passing through the water bear witness to this surface disruption. The most obvious aftermath of passage is the wake, which moves in a predictably angled fashion behind the ship; you can feel those waves if you sit in a rowboat nearby. The subtler aftermath is the change of surface structure — the long, trail-like print left behind by the passing ship (**Fig. 16.7**).

I hadn't noticed this phenomenon until my colleague Michael Raghunath pointed it out to me. Now I see it consistently. The trail typically seems calmer than the water beyond. This calmness makes sense if the ship has disrupted the subsurface EZ structure; the material discontinuity would impair the water's capacity to sustain waves.

Fig. 16.7 *Ferry in Puget Sound approaching Seattle. Note the long-lasting trail behind.* Courtesy, Michael Raghunath.

Flatness may persist long after the ship has passed, often for as much as 15 to 30 minutes; it then disappears. The time required for its disappearance presumably represents the time required for the surface to restructure and become indistinguishable from the rest.

So the water surface layer may be globally elastic but locally fragile — it can break up at any given place. The susceptibility to breakup is created by thixotropy — hard to pronounce but significant for understanding what happens at the water surface.

Cruise Ships, Bathtubs, and Archimedes

All of this brings us to Archimedes, who long ago considered water surfaces. Immersing his mass into a bathtub and watching the water rise, Archimedes had an epiphany: on partially submerged objects, he reasoned, the upward force must equal the weight of the water displaced. It's a simple principle, used to this day to explain why ships float. On the other hand, there may be more to this principle than meets the eye.

First, think of a model ship sitting on a wet sponge. The balance of forces is simple: the boat pushes down, while the slightly dented base pushes up with an equal and opposite force (**Fig. 16.8**). The reason the sponge can push up is that its molecules manage to hold together despite the denting force; molecular cohesion allows it to push back.

Now set the ship on the water instead of the sponge. The force balance ought to be similar: the boat pushing down, and the water

Fig. 16.8 *Cohesion prevents the boat from sinking.*

pushing up. But how exactly does the water push up? If the water molecules beneath were not cohesively linked, then the weight of the ship might split the water like Moses split the Red Sea; the boat would quickly sink. We ordinarily speak of pressure pushing up; however, molecular cohesiveness must enter into the picture.

Readers familiar with the physics of flotation will recall that the standard explanation does not involve cohesiveness; it involves pressure. The weight of the water above creates pressure below. The pressure pushes in every direction, including upward. The farther down you go, the higher that pressure. Therefore, if a boat were to sink deeper, it would experience higher pressure — until the upward-pushing pressure force balances the ship's weight. There the boat should happily sit. Nothing more is required; no mention of cohesiveness.

Do pressure and cohesiveness lead to different explanations? The pressure-based explanation considers the water pressure to exert the same force in all directions. This rests on the presumption that the water's physical properties are the same in all directions. However, that's not necessarily true: the ship's shear may easily break through the mosaic layers, but the moderately distorted mosaic layers beneath and alongside the ship may remain intact because of their high cohesiveness. If so, then the ship will be cradled by an elastic mosaic (**Fig. 16.9**).

Largely intact, this strained elastic mosaic should create upward thrust, pushing up in much the same way as a trampoline. This pushback may help keep the ship afloat. Should bubbling disturb the mosaic, the ship ought to sink lower (see below).

Thus, Archimedes may have been only partially correct. Of course the pressure force pushes the boat upward. However, the magnitude of the pushing force rests not only on the depth and cohesiveness of the water, but also on the cohesiveness of the EZ net lying below the boat's hull. To the extent that the net remains intact though strained, that net may provide upward thrust.

EZ surface structures may help us understand not only how ships float but also how they might sink. Certain regions are

Fig. 16.9 *EZ layers provide cohesion and upward thrust, helping to keep the ship afloat.*

notorious for ships that mysteriously sink *en masse*. The Bermuda Triangle is the most famous (and perhaps most controversial), but others have been described (**Fig. 16.10**). During certain epochs, many ships have mysteriously sunk in those particular zones, far more often than can be explained by chance.[5] Following the most notorious Bermuda Triangle losses, military surveillance records indicated a paradox: the ship debris usually recoverable after a ship disaster could not be found — nothing at all — despite extensive searches, implying that the ships might have plummeted to the ocean floor *in toto* — the hapless ships apparently went straight down.

Undersea discharges offer a possible explanation for ships sinking in this way. Thermal vents and methane deposits become periodically active, releasing bubbles that could easily disrupt the fragile surface structure. Indeed, pilots searching for the mysteriously lost ships have reported locally bizarre-looking surfaces. One tugboat captain who barely escaped sinking described the surface as foamy and choppy, even though the surrounding sea remained perfectly flat.[w3] Thus, the surface seems to play some role. In the same way as underwater bubbling allows high divers to enter the water easily, natural bubbling might just as effectively facilitate the sinking of ships.

Intrigued by the extensive documentation of these losses, several researchers have attempted to determine experimentally whether bubbling could cause sinking. It seems that it can: an amusing video demonstrates sinking in a small chamber,[w4] while a more serious video, recorded by the BBC, documents a speedboat sinking in shallow waters.[w5] So bubbling from beneath can evidently sink boats, although confirmation from other quarters would certainly be welcome.

Capillary Action

Place a wet teabag on a paper napkin and observe. Before long, much of the napkin will become wet. The water can even climb up napkins suspended vertically. Such phenomena are often referred to as "capillary action."

A more standard demonstration of capillary action usually involves narrow tubes known, oddly enough, as "capillaries." When a quartz

Fig. 16.10 *Caught mysteriously in the clutches of the Bermuda Triangle? (Artist rendering.)*

capillary tube is inserted vertically into a container of water, the water inside the tube will quickly rise to a higher level than that of the surrounding water (**Fig. 16.11**). The water seems to defy gravity.

Classical explanations don't help much. They focus on the end result — the upper meniscus — and not on the rise itself. The meniscus is presumed to stick onto the capillary wall, weighed down by the column of water hanging beneath. That hanging load confers curvature onto the meniscus. To achieve force balance, the upward component of meniscus tension equals the weight of the column (**Fig. 16.12**).

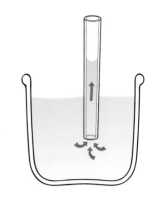

Fig. 16.11 *Example of capillary action.*

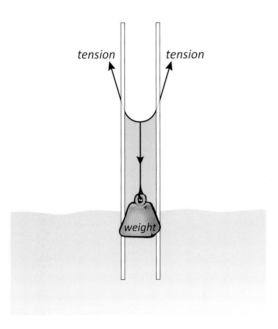

Fig. 16.12 *Balance of forces in the conventional formulation. The upward pull of surface tension balances the weight of the water column suspended beneath.*

That explanation implicitly presumes that the suspended column of water hangs without interacting with the surrounding wall. We know that the hydrophilic walls of these tubes interact strongly with water. So the explanation cannot be entirely correct. Nevertheless, this framework has proved convenient and enduring as an explanatory expedient; students love it.

The classical framework fails to address the more basic issue: why does the column rise? The driving force is left unspecified, although the idea of "surface energy" is often vaguely invoked — i.e., some kind of interaction with the capillary wall.

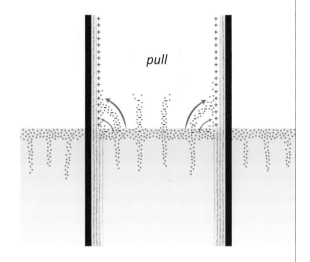

pull

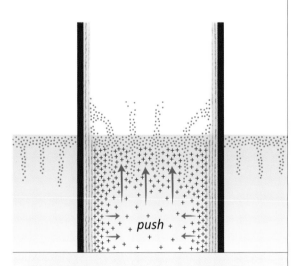

push

Fig. 16.13 *Proposed mechanism of capillary rise. Positive charges from above pull negative surface layer upward (top); and positive charges from below push upward on hydronium ions concentrated immediately beneath the surface layer. Both forces may drive the rise of water.*

Why Does Water Rise in Capillary Tubes?

To help identify the upward-driving force, it helps to begin with the water's functional anatomy. Just beneath the water's surface lies the mosaic-like vesicle array. Since that mosaic is electrically negative, a charged-based rise force seems a good bet — either *pulling* from above or *pushing* from below. Good reasons exist for suggesting both.

First, consider the section of capillary tube projecting upward from the water's surface (**Fig. 16.13**). If EZ layers were to line the tube's inner wall, then those EZ layers should create protons facing the tube's core. Those protons' positive charge could help draw any negatively charged vesicles upward.

Capillary tube walls exposed to the air should bear at least some EZ layers. All hydrophilic surfaces attract airborne moisture. For example, dry substances such as common table salt moisten when exposed to air; some of those substances draw so strongly that they may suffer overnight liquefaction (deliquescence) even in fairly dry environments — including our own laboratory. A particularly avid moisture absorber, Nafion, is marketed as a desiccant. The more hydrophilic the substance, the more atmospheric moisture it will adsorb.

The moisture in the air comes in the form of vesicles (Chapter 15). Vesicles adsorb onto the hydrophilic tube (on both inner and outer surfaces), zippering in the usual way (**Fig. 14.9**). Complete zippering creates standard EZs plus protons. With even a few protons lining the tube's inner surface, the positive charge needed for initiating the upward draw should be present. The walls can begin drawing surface vesicles upward.

Once that upward draw begins, its force should strengthen. As rising vesicles cling to those inner wall protons, zippering will expose more positive charges. Those charges will further enhance the drawing force, pulling up more vesicles, which will then zipper to create still more positive charges, etc. So once the water begins to rise in earnest, it should proceed to completion — when the downward pull of gravity on the water permits no additional rise. You might say the water has pulled itself up in the tube.

The draw of the positive charge may be thought of as an evaporation enhancer. Evaporative events normally occur when the mosaic's local negative charge increases above a threshold; then part of the mosaic array breaks free and rises (Chapter 15). Placing positive charge above that negative array merely accelerates the rise — the positive charge effectively pulling the evaporating water upward. We have confirmed that a positively charged electrode positioned above water can indeed enhance evaporation. So the attractive pull principle seems well founded.

Meanwhile, an upward push may help the rise. The upward push comes from the positive charges beneath.

Consider the section of tube lying below the water surface (**Fig. 16.13,** *lower panel*). Freshly immersed in water, the hydrophilic capillary tube will quickly develop annular EZs, inside and outside. The inside hydronium ions should build in high concentration because of the confining geometry. Those hydronium ions will cling to any negatively charged sites: the annular EZ itself, and the vesicle mosaic hanging from the interface just above. Those two closely spaced and substantial clusters of positive charge will repel. That repulsive force should push the water surface upward.

Thus, the upward driving forces are at least twofold. *An electrostatic force from above exerts an upward pull, while an electrostatic force from below exerts an upward push.* Both forces originate from the positive charges produced from the capillary's EZs.

Do these charge-based mechanisms have explanatory power? Newton apparently thought so — Isaac Newton opined long ago that capillary action might have an electrical origin.[6] His long-forgotten view now seems freshly relevant. The lingering question: can such electrostatic mechanisms account for the basic features of capillary action?

If so, we might first predict that the capillary rise will be most prominent near the wall, where the forces originate. In tubes with large diameters, substantial rise may be evident only right next to the wall, creating an edge meniscus; the rest of the surface may remain flat. In narrow tubes, we may see the rise over the tube's entire cross-section. Those expectations are amply confirmed: near-wall menisci are present irrespective of tube diameter, while full-column rise is seen only in narrow tubes.

A related expectation is that narrower tubes will produce a greater rise. The rise ceases when the downward (weight) force grows high enough to balance the upward force. Those two forces change in characteristic ways: The upward force increases with the capillary perimeter, while the downward weight force increases with capillary cross section. Narrowing the tube diminishes the perimeter and rise force, but it diminishes the cross-section proportionately more. Thus, narrower columns should lift higher because of their substantially diminished weight. That higher lift is borne out by common experience.

A second expectation is that warm water will rise faster. Heating promotes evaporation, and higher evaporation rate should produce a faster rise. We confirmed that warm water rises two to three times faster than room temperature water.

A third expectation relates to the pushing mechanism. That mechanism posits highly concentrated protons in the zone immediately beneath the surface mosaic. Protons in high concentration generate strong infrared signals — we've seen many examples in these pages. We found the same here: strong infrared signals appear consistently just beneath the surface meniscus (**Fig. 16.14**).

A fourth expectation is the absence of any rise in hydrophobic tubes. The driving forces under consideration all stem from positive charges created from EZ buildup. Hydrophobic surfaces produce no EZ; hence water should not rise at all in hydrophobic tubes, and indeed it does not.

A fifth point is the mechanism's apparent universality: the mechanism proposed here is essentially similar to the one advanced to draw water osmotically (**Fig. 11.8**) and the one advanced to draw water

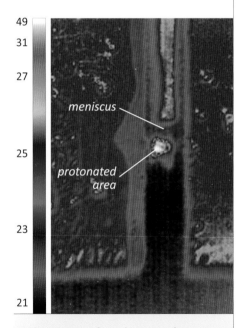

Fig. 16.14 *Infrared image of water rising in a square capillary tube. Note the hot spot just beneath the meniscus. Scale in °C.*

along mesh-like substances (**Fig. 11.11**). All are proton driven. Hence charge-based mechanisms can govern a multitude of water-drawing phenomena — perhaps all of them. This universality is a strong point for the electrostatic mechanism.

Water Transport in Tall Trees

Capillary action is not restricted to quartz tubes and paper napkins alone. It occurs throughout nature. Capillarity is especially prominent in the plant kingdom, where water may rise even to the tops of redwood trees 100 meters tall. Inside such trees, narrow xylem vessels run from roots to leaves, transporting the water ever upward.

The mechanism of vessel transport is actively debated. Many scientists think that some kind of capillary action draws the water upward. However, two issues plague that hypothesis. First, the "hanging column" is too heavy to be lifted more than about 10 meters; second, the air pockets commonly found within the fluid of xylem tubes should thwart the upward drawing process, just as they do in straws. Scientists struggle with those issues.

In the capillary mechanism outlined above, these issues are not necessarily impediments. The column does not hang — it clings to the walls; and air pockets need not impede the charge-based rise. The question is whether the mechanism really does operate in plants and trees. A key issue is whether xylem tubes contain exclusion zones, and the answer appears to be yes.

To find out about EZs in xylem tubes, I contacted my Australian friend Martin Canny, the dean of the plant vascular field. Martin lives in Canberra. I recall a visit several years ago, when the Cannys were kind enough to put me up in their mother-in-law apartment downstairs. Martin mentioned the spiders. No need to worry about the Huntsman spider, he indicated dismissively — it's huge and hairy but perfectly harmless. But do watch out for the little black ones with red spots. They lurk in nooks and crannies. Death from their venom may be quick enough, but it's also agonizing. Needless to say, my three-day visit went practically sleepless.

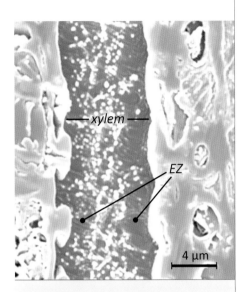

Fig. 16.15 *Cryo scanning electron microscope image of ink particles infused into xylem. Particles concentrate in the center and are excluded from the zone near the walls.* Courtesy Martin Canny.

Nevertheless, we did have a chance to discuss capillary action, and Martin seemed curious about exclusion zones. Following my visit, he went ahead and checked. He infused small ink particles into xylem tubes, quickly froze the specimens, and examined the frozen samples in an electron microscope. The results were positive (**Figure 16.15**). I'm not sure who was more excited, Martin or me, but the results confirmed the presence of exclusion zones in those vessels.

This confirmation implied that we were on a productive course. If annular EZs are present in xylem tubes, then surely they play some role in the tubes' physiology. Annular EZs in Nafion tubes generate steady intratubular flow (Chapter 7), and intratubular flow is exactly what's needed. Indeed, flow inside the Nafion (or gel) tube could model what happens in nature's vascular tubes.

In the Nafion flow model, a central feature is the presence of core protons. Those protons drive the flow. But do protons also fill the xylem fluid? Standard textbooks confirm the sap's expectedly low pH, and modern methods narrow down those pH values: in maize seedlings, for example, depending on conditions, xylem pH ranges between 5 and 4.[7]

Thus, xylem tubes seem very much like Nafion tubes. Both bear annular EZs, and both contain core protons. To suggest that the drivers of flow are the same requires no giant leap. We may continue to refer to those flows as capillary driven, out of habit, but it might be more accurate to refer to *proton-driven* flow.

Proton-driven flow replaces the water lost through evaporation. Water evaporates from the leaves of the plant. As it does, the top of the xylem vessel may become transiently dry — except for a few residual EZ layers. The protons clinging to those EZ layers would then draw the water upward from below by the same mechanism that draws water upward in narrow quartz capillary tubes. That upward movement keeps the leaves hydrated.

Height should not be a limiting issue for this mechanism because the tubes are sufficiently narrow to make the drawing force stronger than the gravitational force. The upper vessels of trees have micrometer scale diameters. Vessels farther down are wider, but their lumens commonly abound with hydrophilic polymer strands, which effectively narrow the tubes. EZs cling to all of those surfaces, and bulk water clings to the EZs because of its many hydronium ions. These clingy attachments bear much of the column's weight. With the tube's substantial load-bearing capacity and narrowness, water should experience no difficulty rising to great heights.

The energetics of this process seem worthy of comment. The upward flow requires energy, just as pumping water to an elevated storage tank requires energy. The source of energy is familiar: incident radiant energy. In the same way that radiant energy fuels the flow of water inside hydrophilic tubes, radiant energy should similarly fuel the flow through xylem tubes.

Given the direct contribution of incident radiant energy to flow, you can understand why this flow might depend on the season. Flow begins as spring approaches — just when ambient radiant energy begins picking up in earnest. The flow increases as summer approaches, slows down in autumn, and shuts off in winter. The autumn reduction of radiant fuel supply might directly explain the reduction of flow — and why autumn leaves dry out and fall gently to the earth below.

The colors of autumn.

Floating Water Droplets

Finally, we return to consider the surfaces of natural bodies of water, asking what happens when rain falls on those surfaces. Intuition suggests that the drops will instantly coalesce with the body of water below. However, if both the droplet and the water surface bear EZs, then coalescence won't necessarily occur instantaneously.

I first learned of delayed coalescence when a student told me of his experience sailing just after a rainfall. Water would settle onto the boat's gunwales, frequently falling onto the lake below in droplets. As often as not, those droplets would float for some time before dissolving. Once you've seen those floating droplets, you look everywhere for their presence. In rainstorms, they can seem surreal — clear marbles floating on the water's surface.

Delayed droplet coalescence turns out to be a recognized phenomenon, studied on and off for a century. Few people seem aware of it. Confident that further study might divulge more of water's closely guarded secrets, we carried out a detailed investigation of the phenomenon using high-speed video.[8] As long as conditions were properly set and the droplets were released from heights of less than 10 mm above the water surface, the droplets consistently floated before dissolving (**Fig. 16.16**).

Further, when the droplets happened to roll or slide sideways, coalescence took much longer — sometimes many seconds. The delay may result from the additional time needed for breaching the droplet's EZ shell. If the droplet rolls, then the contact point continuously changes, and the breaching process must continuously begin anew. Dissolution therefore requires more time.

Actual dissolution occurs in more than just a single step.[8] The process involves a series of five or six squirts, each one ejecting some of the droplet's contents into the water beneath. Some of those downward squirts are powerful enough to induce waves in the water — and even to drive the residual droplet upward (**Fig. 16.17**, *opposite*).

Some panels of **Figure 16.17** may look familiar. Similar images are often reproduced in books, magazines, and websites. The characteristic

Fig. 16.16 *Water droplets falling onto water. Under appropriate conditions, droplets can persist for some time before coalescing with the water below.*

dance steps remain enigmatic — a challenge for the reader to resolve. Possibly, the initial squirt relieves much of the droplet's internal pressure. If the resealed droplet contains residual positive charge, then the sequence may repeat. Only after multiple squirts will the droplet empty, adding the proverbial drop to the ocean.

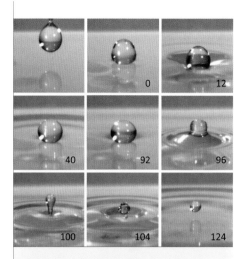

Fig. 16.17 *Several stages of water droplet dissolution in water.*[8] *Numbers indicate time from when the droplet lands, in milliseconds.*

Summary

EZ-containing structures line the water surface. These subsurface structures consist mainly of aggregated vesicles, and possibly also standard EZ material, self-organized into mosaic-like arrays. These arrays may project down from the surface by millimeters or centimeters in laboratory vessels; in open waters with ample incident radiation, they may penetrate tens, or even hundreds, of meters. Those tubular mosaic structures create interfacial tension.

Indeed, interfacial tensions in natural bodies of water may be extremely high — certainly enough to support small lizards and perhaps also capable of assisting in the support of ships. The agent creating all that tension is the subsurface EZ water mosaic. Its presence helps explain many observable phenomena: persisting ship trails, sustained tsunami waves, capillary flows in trees, and the enigma of water droplets floating on water surfaces.

Whether such mosaic structures can help you walk on water is another issue. From this vantage point it seems doubtful — although we're told that it has happened before.

GRANDMA'S "FAMOUS" ICE RECIPE

INGREDIENTS:
- 1 QUART EZ WATER
- PROTONS, AS NEEDED

• POUR EZ WATER INTO A MEDIUM SIZED BOWL. BLEND IN PROTONS AND STIR WELL.
• QUICKLY POUR MIXTURE INTO LARGE ICE TRAY.
• SERVE WHEN WELL CRYSTALLIZED.

17 Warming Up to Ice

In 1963, Erasto Mpemba was a middle-school student in Tangan-yika (now Tanzania). The topic of the day in his cooking class was ice cream. Most Tanganyikan middle-school students were hardly aspiring chefs, so shortcuts were common. To make ice cream, the students would simply dump a premixed powder into water, stir the mixture, and shove it into the freezer. Soon, they could enjoy their treat.

Mpemba noticed something odd. When he mixed his powder with warm water instead of cold water, the ice cream was ready sooner. The warm water seemed to freeze faster. That seemed impressive to Mpemba — though quite improbable to his middle school teachers.

Erasto Mpemba, 1950 -

Mpemba couldn't forget his counterintuitive observation. When he moved on to high school, he mentioned the paradox to his teacher, a physics professor recruited to awaken high-school students to the marvels of science. Denis Osborne was not impressed. Every thread of known thermodynamic fabric led him to conclude that warm water could not possibly freeze faster than cold water. Nevertheless, Mpem-ba persisted. Finally, he persuaded Osborne to try the experiment himself using pure water; to his surprise, Osborne quickly confirmed Mpemba's observation.

Their joint papers are now classic.[1,2] You can watch modern video demonstrations of the phenomenon.[w1,w2] The credit rightly goes to the middle-school student alert enough to sense the importance of his paradoxical observation, unaware that several notables, including Aristotle, had observed similar phenomena much earlier.

The so-called "Mpemba effect" is not the only paradoxical feature of ice formation. Ice may look like a solid, inanimate block of water, but its formation presents several additional paradoxes that will need to be resolved if we are to really understand how ice forms.

We start by asking the now-familiar question: might the EZ play a role? Given the exclusion zone's structural similarity to ice, some involvement of the EZ seems inevitable. For example, the EZ state might precede freezing; it might also follow melting. If so, then we might ask whether a drop in temperature is really the most critical factor for ice formation. Or could cooling merely set the stage for some other ice-forming process?

We begin with an ice formation paradox even more fundamental than Mpemba's.

The Energy Paradox

In order for water to freeze, common experience suggests that energy must be withdrawn. Think of the process. You insert a tray of water into your freezer. When enough heat has been extracted, the water turns to ice. You may then do the reverse: expose it to warm air, and the ice melts. So adding energy yields the disordered liquid, water; withdrawing energy yields the ordered crystal, ice.

While this is familiar territory, something seems curious. According to all we've learned in early chapters, creating crystalline order requires the *addition* of energy; to build order (and reduce entropy) you ordinarily need to put energy in. Building the ordered EZ requires electromagnetic energy — the more energy supplied, the larger the ordered zone.

That feature follows common sense. If you want to create an elaborately structured sand castle, you must expend energy for its construction. Returning the edifice to rubble can require almost no energy — a gentle tap in the right place — but the task of building the ordered structure always demands substantial energy input.

If all of this seems reasonable, then we are in deep trouble, because the freezing scenario seems backwards. Ice is the ultimate ordered crystal. As such, you'd expect that lots of energy should be pumped in to fuel its construction. Yet, common experience seems to tell us that energy needs to be *pulled out*.

Scientists have rationalized the ice-formation anomaly by invoking the notion of thermal motion. Thermal motion diminishes as the temperature descends, so reducing the temperature should allow water molecules to follow their natural tendency to self-organize into crystalline ice. On the surface, that seems reasonable; however, it raises the thermodynamic question: if increasing order to form the EZ requires energy *input*, how can increasing order to form ice require energy *withdrawal*?

My friend Lee Huntsman, a fellow scientist/engineer who defected from working in the scientific trenches to become our university's president, brought this paradox to my attention. Lee approached me after a public lecture that I was invited to deliver at my university. Coming amid the congratulatory nods following the lecture, Lee's question about the thermodynamic paradox was the interaction that I most appreciated. It stirred me to think.

Eventually we resolved this paradox, and in so doing gained fresh understanding. I came to appreciate that a massive reserve of internal energy stands ready to fuel the transition from liquid water to ice. Triggering the release of that energy requires adequate cooling. In the end, energy really does get used to create order — just as we had observed earlier for EZ formation.

Resolving the Energy Paradox: EZ as Protagonist

While ice obviously forms from water, it's less clear whether it forms from bulk water or EZ water. Since EZ and ice structures both comprise stacked honeycomb sheets, it seems natural to suggest a linkage between those phases. An EZ-to-ice transition can be easily envisaged.

The differences between EZ and ice structures offer clues about what might drive such a transition (**Fig. 17.1**). In the EZ structure, the atoms in adjacent planes are offset from one another (*panel a*). Charges of one plane line up with opposite charges of the adjacent plane. Attraction keeps EZ planes stuck together. Ice, by contrast, has its honeycomb sheets stacked in register (*panel c*). Oxygen atoms lie opposite oxygen atoms, and hydrogen atoms lie

(a) EZ structure

(b) protons introduced

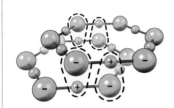

(c) layers align to form ice

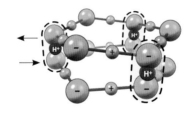

Fig. 17.1 *Transition from EZ* (a) *to ice. The transition requires protons* (b) *and planar shift* (c). *(The ice planes are not flat because of local attractions and repulsions.)*

opposite hydrogen atoms. Their proximity creates local repulsions. Those repulsive forces would ordinarily push apart the hexameric planes, exploding the structure, but nature employs a clever trick: gluing planes together with protons *(panel c)*. The protons insinuate themselves between every other juxtaposed oxygen pair (on adjacent planes). In this way, a positive charge glues two negative atoms. That solidifies the EZ water into ice.

Evidently, any transition from EZ to ice requires a massive influx of protons *(panel b)*. Those protons add positive charge to the EZ lattice, shifting that ordinarily negative structure towards the net electrical neutrality of ice. The protons also take up space: their presence pushes the EZ planes apart, accounting for ice's lower density (enabling it to float). Thus, an EZ-to-ice transition seems a promising option; it can explain at least some of ice's basic features.

An EZ-to-ice transition also makes sense energetically. EZs exclude protons. Those excluded protons constitute potential energy: positive charges separated from negative. If, at some point in time, protons rush back in — neutralizing the negative EZ and thereby creating ice — potential energy gets surrendered. Therefore, our expectations are satisfied: energy gets used for converting an ordered structure into an even more ordered structure (**Fig. 17.2**).

Fig. 17.2 *Energetic aspects of the transition from EZ water to ice. The required energy comes from separated protons.*

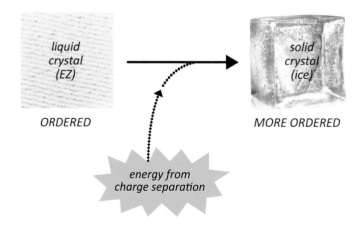

liquid crystal (EZ)

solid crystal (ice)

ORDERED

MORE ORDERED

energy from charge separation

Given these sensible features, we proceeded to test whether the EZ-to-ice transition might be more than just a theoretical nicety; we examined exactly how ice forms.

Evidence that an Exclusion Zone Precedes Ice

To carry out these experiments, we used a large cooling plate. Atop the plate, we positioned a strip of Nafion, next to which we dropped some water (**Fig. 17.3,** *diagram*). We then began cooling the plate.

The near-Nafion EZ was always the first region to freeze, well before any bulk water. Typically, the freeze began at some point along the water-Nafion interface (*left panel,* white spot). Freezing would then propagate, usually faster along the EZ than perpendicular to it; sometimes, long stretches of EZ froze before any of the bulk water just beyond (*right panel,* arrow).

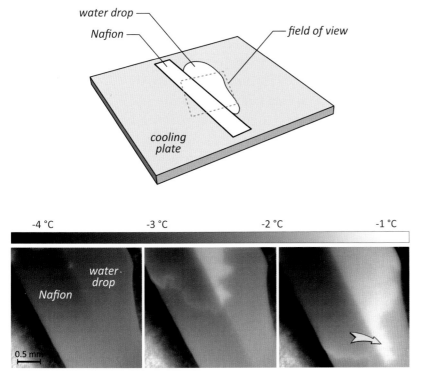

Fig. 17.3 *Sequential progression of droplet freezing on a cold plate, as recorded by an infrared camera. The droplet sits beside a strip of Nafion. Region of freezing begins in the EZ next to Nafion (left, white dot) and tends to propagate most rapidly along the EZ (arrow, right panel).*

We found much the same EZ freeze in an entirely different experimental setup (**Fig. 17.4**). We inserted a tongue of heat-conducting material into a small experimental chamber containing water and microspheres. The tongue could be cooled from the outside, either by a thermoelectric device or by exposure to a fluid coolant. Either way, the tongue drew heat from the water.

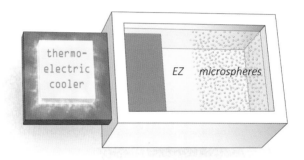

Fig. 17.4 *Cooling apparatus for exploring the relation between EZ and ice formation.*

As the water began cooling, nothing much happened at first. Then an exclusion zone began to form next to the tongue. (The tongue's material didn't affect this process much, as long as it could conduct heat.) As cooling proceeded, the EZ grew progressively larger — often to 500 micrometers or more. Then, either of two things happened: In some instances, the microspheres suddenly invaded the EZ, bunched up, and froze. We knew the EZs had frozen because the microspheres they enclosed were utterly crushed. In other instances, the microspheres remained excluded, and the original EZ froze. Either way, the EZ was the ice precursor, just as it was in the Nafion-droplet experiment.

A side question: why does the cooling tongue generate an EZ? This apparent paradox puzzled us at first because EZ buildup generally requires infrared energy input. If the cooling surface withdraws infrared, we would expect any incipient EZ to shrink, not expand.

We finally deduced that, in this configuration, the EZ receives a lot more infrared than anticipated because of an asymmetry. In the setup of **Figure 17.4**, the cold plate certainly draws infrared leftward, out of the adjacent water. However, the bulk water beyond also pumps IR toward the frigid plate; that IR passes *into* (thereby helping to build) the incipient exclusion zone. This push-pull scenario creates a significant infrared throughput: IR energy flows in abundance through the EZ; therefore, the EZ expands. It is this expanded EZ that evidently turns into ice.

This answer to the side question seemed satisfying because EZ growth meant that protons ought to accumulate beyond the growing EZ. Those separated protons would then be poised for their hypothesized invasion into the EZ. We felt we were on a roll.

What about melting? If freezing involves a transition from EZ to ice, then melting should involve a transition from ice to EZ. That expectation was easily confirmed.[3] We placed small blocks of ice into standard spectrophotometer cuvettes and probed the melting water. Consistently, the 270-nm EZ-signature peak appeared as the ice began melting. The peak showed up no matter what type of water had been frozen (**Fig. 17.5**). The peak persisted for some tens of seconds and then vanished as the melting ice completed its transition to bulk water.

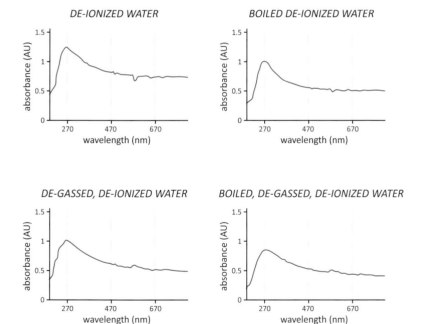

Fig. 17.5 *Spectrometer readings obtained from just-melted water. The 270 nm peak is consistently present.*

So it didn't matter whether water was freezing or melting. Either way, the EZ and ice states were intimately linked. No surprise, considering their similar structures.

Temperature and Ice Formation

Having confirmed the EZ-ice linkage, we wanted to explore the second feature of the hypothesis: proton influx. Those protons seemed readily available — produced by EZ buildup and available to invade the EZ to create the ice.

But first we needed to address the issue of temperature. Most scientists agree with the common view that cooling, not proton influx, is the critical factor for ice formation. Asserting the primacy of proton influx would make cooling secondary, which seemed weird: how could cooling play a supporting role if freezing always occurs at a fixed temperature? So well recognized is that fixed freezing temperature that it even serves to define a benchmark on the Celsius scale.

While the 0 °C freezing temperature may be common knowledge even to children, it turns out that water does not *always* freeze at that temperature. I don't mean water with dissolved matter, where so-called "colligative" properties may lower the critical temperature by a few degrees. I'm referring to pure water under standard conditions, where the freezing temperature can sometimes descend far below 0 °C.

Scientists recognize that pure water at atmospheric pressure at sea level sometimes does not freeze until the temperature descends to nearly -40 °C. In confined spaces, freezing temperatures can descend further, even to as low -80 °C.[4] Difficult-to-freeze water is well recognized but poorly understood. Such water is expediently classified as "super-cooled," a label that merely distracts attention from the absence of real understanding,

That the threshold temperature for freezing can descend so low seems fortunate. If water always froze at 0 °C, then plant life would be extinguished in colder climes: each plant's water would turn to ice, which would rip mercilessly through the entire plant, shredding even its organelles. That does not happen because the freezing temperature can descend to values much lower than the frigid ambient temperature.

On the other hand, learning that temperature is not quite the decisive variable we might casually suppose should seem less surprising in light of the term's inherent imprecision (see Chapter 10). The ambiguity of "temperature" led us to seek another, more meaningful variable to understand how ice forms. That variable, according to the hypothesis under consideration, should relate to the invading protons.

An Israeli group recently shed light on this issue by investigating the role of charge during freezing.[5] The investigators immersed a pyroelectric device in an externally cooled chamber (**Fig. 17.6**). Pyroelectric devices can control their surface charge polarity. This allowed the researchers to test how negative, neutral, or positive surface charges affected freezing.

Negative-charge polarity made it difficult to freeze the contiguous water (*top*); i.e., water temperature had to reach very low values before the water could freeze. Positive polarity did the opposite — it promoted freezing (*bottom*). That is, when the surface presented a positive charge, the contiguous water could freeze before its temperature descended much. Positive charge promoted freezing.

A notable aspect of that experiment was the order of freezing. The water-filled chamber was being cooled from all around; ice should have begun forming at the periphery of the chamber. Generally, ice did form there, but not during the positive charge protocol. In that case, the ice began forming right at the pyroelectrically charged interface, implying the importance of positive charge for ice formation. Even more telling: to generate that positivity, the pyroelectric device had to be *heated*. In other words, the water froze preferentially next to the device even as

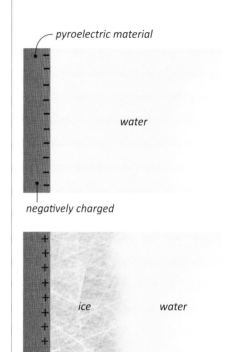

Fig. 17.6 *Pyroelectric experiments. Negative charge retards ice formation, while positive charge promotes ice formation.*

ROOM TEMPERATURE

BEGINNING TO FREEZE

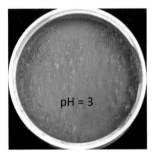

Fig. 17.7 *Water containing diluted pH dye. pH values refer to the chambers' peripheries.*

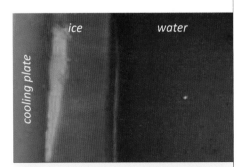

Fig. 17.8 *Color of pH-sensitive dye during freezing. Deep orange color indicates low pH and hence abundant protons.*

that device was *adding* heat to the water. So much for cooling as a requirement for ice formation!

This latter result makes it crystal clear — even if earlier paragraphs did not — that temperature cannot be the decisive variable for freezing. Depending on circumstances, even *adding* heat can create ice.

Those experimental results show the importance of positive charge for ice formation. They also support our hypothesis, since the positively charged protons could, if they invaded the EZ, likewise facilitate ice formation. The experiment makes those protons seem at least as important as any lowering of temperature.

The Proton Rush

Buoyed by these observations, we proceeded to test whether protons actually invade as ice forms. We first repeated early observations — half-century old classical electrical measurements that implied the appearance of positive charge during freezing.[6] We placed an electrode ahead of an advancing ice front. As the ice front approached, the electrical potential spiked upward by as much as one volt. This confirmation of an increase in positive charge provided impetus to check further using direct visualization.

To do so, we used pH-sensitive dye (see Chapter 5). In one experimental setup, we used a circular chamber (**Fig. 17.7**). We placed that chamber atop a liquid-nitrogen-cooled metal plate. At room temperature, the dye showed green (*top*), indicating neutrality. As the water began freezing at the rim, the rim color turned deep orange (*bottom*), indicating numerous protons in the regions undergoing freezing. These experiments could not reveal how those protons got there; but only that they entered the region of ice formation, as anticipated.

We got a similar result when we inserted an aluminum cooling plate into the chamber of water. Again, the area where the ice formed turned deep orange, indicating the presence of many protons (**Fig. 17.8**).

drop

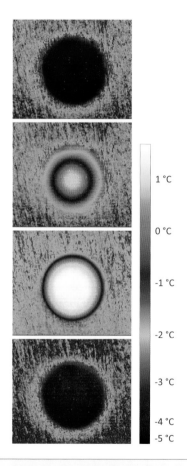

Fig. 17.9 *Sequence of color change during freezing of a droplet on a cooled surface. Orange color shows the concentration of near-surface protons as the drop began to freeze.*

We obtained another, similar set of results using droplets. A droplet, placed atop a cool plate, froze from its base upward. We could examine the periphery of the droplet, not the inside. As the periphery froze, its color shifted from green to orange, once again indicating incoming protons (**Fig. 17.9**).

Confident that the results were confirming the anticipated proton rush, we pursued yet another experimental approach, using an infrared camera. Here again, we examined a droplet on a cooling surface. If protons flooded the droplet's peripheral EZ, then that movement of charge should create a flash of infrared energy. Charge movement generates infrared (Chapter 10); we've already seen many examples. We confirmed the infrared flash — the burst of IR persisted for a second or so (**Fig. 17.10**). The presence of that flash lent further support to the hypothesis of a proton rush, and also left us appreciative that the droplet could signal its transition to ice in so "brilliant" a way.

Our excitement over that brilliance was briefly dashed when somebody suggested a different interpretation, that the flash merely signaled the "heat of fusion." Water supposedly loses heat as it freezes; the expression of that heat loss might be the infrared flash. That interpretation seemed plausible at first; however, something didn't fit. Withdrawing heat should cause the droplet to cool, or at least remain at the same temperature; but the temperature scale (**Fig. 17.10**, *right side*) seems to indicate that the droplet heated as it froze (*third panel*). Why should something undergo a temperature *increase* during the course of freezing? That made no sense at all.

The infrared flash made a lot more sense when interpreted as a signal of rapid proton influx. The charge movement created the flash. That interpretation complements the pH dye results, which more directly confirm the rush of protons. Hence, the two sets of results reinforce one another: protons do evidently rush in to create the ice.

1 °C

0 °C

-1 °C

-2 °C

-3 °C

-4 °C

-5 °C

Fig. 17.10 *Infrared emission from a water droplet during freezing on a cold surface. The droplet emits a brief flash of infrared light as it freezes* (top to bottom). *Frame times in seconds: 3.3, 29.0, 29.3, 30.0. Equivalent temperature scale is shown at right.*

The Proton-Release Trigger

If protons play the role anticipated, then some trigger must unleash the excluded protons to invade the EZ. Without a release trigger, protons might seep continuously back into the EZ, creating ice all the time. That doesn't happen.

What might trigger such an invasion?

Recall that ice crystal formation requires bare protons; those protons insinuate themselves between the EZ planes to create the ice (**Fig. 17.1**). However, bare protons are generally unavailable. The usual proton-containing species is the hydronium ion — a proton attached to a water molecule.

You'd think that hydronium ions might suffice, but size is an issue. Hydronium ions may draw strongly toward the negative EZ lattice, but the ion's bulk prevents entry (**Fig. 17.11**). Even water molecules are too large to enter. The magnitude of the obstacle becomes clear from considering the size of the portals running through the EZ. The unit hexagons are miniscule. Even more serious, the hexagons of adjacent planes lie out of register (Chapter 4); this reduces the openings to virtually nothing. Only the diminutive proton is small enough to penetrate.

Fig. 17.11 *Only protons are small enough to penetrate into the EZ lattice.*

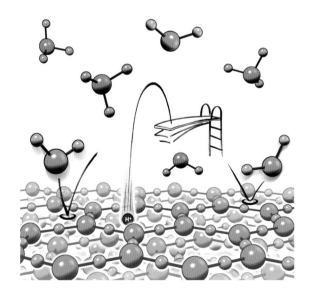

Protons, however, are ordinarily unavailable. To become available, protons must free themselves from their parent hydronium ions. Only then can they rush into the EZ lattice to create ice. Thus, proton release may be a possible trigger for ice formation. Whatever agent can free the proton from the hydronium ion could trigger the formation of ice.

To identify that agent, think of the forces exerted on the proton. One force comes directly from the EZ's negative charge, which pulls on the positive proton as if to suck it into the EZ (**Fig. 17.12**). That pulling force lacks enough strength to dislodge the proton; otherwise, since the force is always present, protons would continuously enter the EZ, defeating any prospect of maintaining bulk water's observed positivity. The EZ's pull may help the proton gain its freedom, but it does not suffice.

A second force is the push. This comes from all of the hydronium ions lying beyond (in the figure, *rightward of*) the vulnerable hydronium ion (**Fig. 17.12**). Their positive charges repel the proton in question. This push force acts to dislodge the proton, whereupon the negative EZ can pull the proton all the way inward.

To dislodge the proton, the push force must exceed some threshold. The force's magnitude will depend on how many hydronium ions are pushing, which in turn depends on the number and distribution of hydronium ions. Once the number of hydronium ions grows large enough — if it does — near-EZ protons may begin breaking free. Those protons should then get instantly sucked into the EZ, penetrating all the way to the deepest layer with most negative charge. That event should initiate ice formation.

Cooperative Ice Buildup: Why Ice Is Inevitably Solid

The mechanism outlined above should explain the creation of ice: An EZ grows, while hydronium ions build up beyond the EZ. Protons dislodged from those hydronium ions penetrate all the way to the most negative (i.e., deepest) EZ layer, then to the next, and so on. Ice builds progressively.

Everything seems in order — you'd think that ice could grow by this mechanism into a solid mass. However, any quirk that leaves even a single layer unprotonated could wreak havoc — two small blocks of

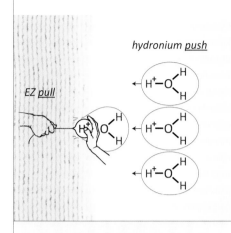

Fig. 17.12 *Forces exerted on a near-EZ hydronium ion. EZ negativity pulls on the proton* (left) *while many bulk-water hydronium ions push* (right)*, eventually dislodging it from the parent water molecule. Once dislodged, the proton can easily penetrate into the negative EZ.*

ice might form instead of a single large one. Since that doesn't happen, nature likely employs some feature that ensures ice's integrity: e.g., a just-formed layer of ice could facilitate the formation of the next layer. Cooperativity of that kind could ensure that ice cubes remain solid.

Cooperativity turns out to be an inherent feature of the proposed proton-invasion mechanism. Envision two adjacent planes (**Fig. 17.13**). In the EZ state, those planes lie out of register (*left*); in the ice state, they lie in register (*middle*). In order for ice to form, plane *B* must shift relative to plane *A*.

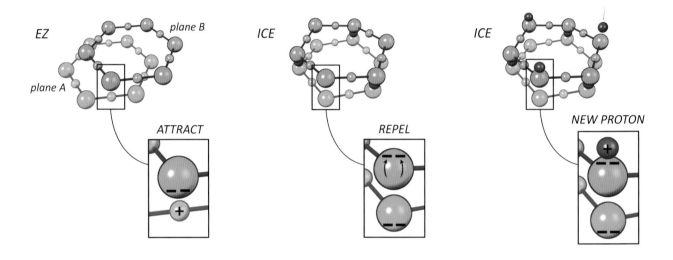

Fig. 17.13 *Cooperative nature of ice formation. EZ configuration shown at left. In newly created ice layer* (middle)*, oxygen's negative charge shifts upward (box), attracting incoming protons. This shift occurs in the three planar oxygen atoms juxtaposed to oxygen atoms beneath. Incoming protons settle at those three points* (right)*, setting the stage for the next ice layer.*

When envisioning this shift, consider the positions of oxygen's electrons. In the EZ configuration shown in the figure, plane *B*'s oxygen electrons orient toward plane *A*, attracted by the nearby positive charge (*left, box*). That attraction keeps the planes stuck together. Once plane *B* shifts to the ice configuration, the environment changes. An oxygen atom replaces the lower hydrogen (*middle*). The adjacent oxygen atoms that are not glued by incoming protons now repel, shifting their electrons to opposite directions (*middle, box*). The electrons of the upper oxygen now face the next wave of protons invading from above, providing an attractor (*right*). You might say that the shift puts out the welcome mat for the invading protons.

That welcome mat lies at the face of the just-forming ice. Its presence assures that the invading protons stick where they should to form the next layer. In this way, the ice assuredly builds plane by plane, uninterrupted. The ice is certain to be solid.

A satisfying feature of growing the ice layers as proposed is that they are correctly structured. Please note the locations of the protons between layers (**Fig. 17.13**, *right panel,* dark blue dots); the protons link every other oxygen. Those protons are 60° offset from the protons of the next layer. That regular shift of 60° per plane yields the correct ice structure.

So the model of ice buildup seems on track. The model's cooperative nature ensures that ice will be solid; and the model's detailed operative features ensure that all protons will be correctly positioned.

Natural Ice Formation

To understand how ice builds in natural settings according to the proposed model, consider the freezing of a lake. Say a wintry cold front arrives. The air above the lake imparts an icy chill. Evaporation diminishes. Surface EZs remain largely in place, forming a relatively stable EZ cap — something like a lid placed on a container of water (**Fig. 17.14a**).

That EZ cap radiates infrared to the colder air above; meanwhile the cap *receives* IR from the warmer water beneath (**Fig. 17.14**). This large infrared throughput permits the EZ to grow — in a manner similar to the scenario of **Figure 17.4**. The hydronium ions accumulating beneath cannot escape — the cap confines them. Once their concentration grows enough to cross the critical threshold, dislodged protons begin invading the EZ, penetrating all the way to the most negative region at the top. Those protons build the topmost layer of ice (**Fig. 17.14b**).

Meanwhile, the EZ continues to grow because of the sustained infrared throughput. In this way, the ice thickens. Cooperativity assures uniform ice buildup. The colder the air, the thicker the ice. When the ice thickens enough to diminish the vertical infrared gradient, this finally lets the ice stabilize.

Thus, the ice-formation principles operating in laboratory chambers appear to operate similarly in nature.

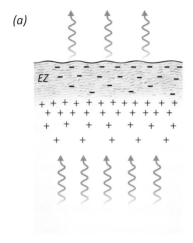

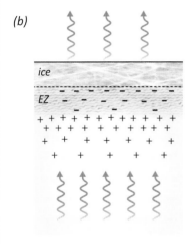

Fig. 17.14 *Ice formation in natural bodies of water. Infrared energy from the warmer water beneath builds the surface EZ and separates charge* (a). *Invading protons create ice* (b). *Ice thickens as the process continues.*

Freezing from the Surface Downward

Top-down freezing is a fortunate natural phenomenon. If freezing happened from the bottom up, then not only would ice-skating be awkward, but also the fish nudged out of the frozen water might find themselves gasping for dear life. Fortunately, ice always grows from the surface downward. Fish remain happy.

Energetics Resolved

With this understanding in place, we can now deal with the issues of energetics. Two questions arose. The first asked why ice-formation energetics seemed to deviate from expectations. That question has been resolved. Ice formation *does* require energy; it exploits the potential energy of charge separation which is delivered as the positive proton charges combine with the negative EZ to build crystalline ice. Hence the energetics of ice-formation match the energetics of EZ buildup. Both processes require energy.

Requiring energy for creating order is a principle that may confuse. For at least one kind of energy — heat — conventional thermodynamics argues the opposite: heat increases "thermal motion," thereby promoting *disorder*. In the era of steam engines when thermodynamics was formally conceived, heating a vessel of water to steam seemed surely to increase randomness. However, Chapter 15 showed evidence of more order as radiant energy drove water toward steam, not less order. Thus, one of classical thermodynamics' central tenets has come open to question. Does heat energy really lead to disorder?

This question opens another Pandora's box: energy *type*. Thermodynamic principles grew from considerations of heat, and were later extrapolated to apply to all forms of energy. That extrapolation contains an element of self-contradiction: heat energy has been presumed to create *disorder*, while more generally the input of energy is needed to decrease entropy and create *order*. The latter principle squares with common experience and all we've seen in these pages; the former might be erroneous (Chapter 15). Possibly, all energy input creates some kind of order, including the ordering of electrical charges.

Energetic questions of similar nature arise with respect to the formation of salt and sugar crystals (Chapter 10). There again, crystallization takes place as solutions cool (**Fig. 17.15**); hence, it would appear that the ordering occurs as energy is being withdrawn. However, there is a catch: the solutions are generally *heated* beforehand; if not, the required infrared energy may be absorbed from outside. That absorbed energy creates EZs and separates charge. The separated charge provides the energy that drives the ordered crystal formation, here by the like-likes-like attraction (Chapter 8). Thus, the energetics powering the formation of salt and sugar crystals may operate much like the energetics involved in the formation of EZ and ice. Energy drives ordering.

This same principle may also hold for metals. Common metals exhibit atomic crystallinity. When heated to melting, metals become more amorphous. The question arises whether the radiant input used for melting provides the energy needed to fuel recrystallization, as in salt and sugar crystals. If so, then the same thermodynamic principle could remain applicable: ordering requires energy.

Returning to ice, let us now consider the second of ice's seemingly anomalous energetic features: the so-called latent heat. As commonly understood, latent heat is the heat given off as water transitions into ice. The surrendered heat is believed to warm up the immediate environment, while the water itself is presumed to remain at constant temperature. However, that's not what we observed. Infrared images of freezing droplets exhibit hardly any heating of the surrounding area, while the water being frozen "heats up" during the transition to ice (**Fig. 17.10**). A conventional interpretation of these IR images would assert that the water *gets hotter* as it transitions to ice. That's not supposed to happen.

A more plausible interpretation of the "latent heat" follows from our earlier understanding that the infrared radiation originates from charge movement. The proton rush is exactly that — a massive movement of charge. Hence, the so-called latent heat may be nothing more than an expression of that proton rush. The "heat" inferred from the infrared flash has no intrinsic significance.

Fig. 17.15 *Rock candy. Immersed strings serve as nucleation sites for sugar crystal growth. The sugar solution is first heated; then, as the solution cools, the crystals form.*

Mechanical Perturbations Can Trigger Ice Formation

Some eye-popping scenarios confirm the ice trigger's abrupt, threshold-like nature.

Insert a sealed flask of water into your freezer. When the water is super-cooled but not yet frozen, remove the flask. A vigorous shake or a strike on the table can trigger a sudden, massive ice formation.[w3]

Those mechanical perturbations may act by causing protons to shear from their parent water molecules; the protons can then invade local EZs to initiate ice formation. Shaking may also facilitate matters by creating bubbles, which comprise EZs and positive charge — the two essentials needed for freezing. Mechanically induced bubble fracture from striking or shaking could thus flood the system with the essentials need to initiate the freeze.[w4]

Another notable example: remove another almost frozen flask of water from your freezer. Unscrew the cap, and pour the nearly frozen water into a beaker of cold water from a height of about 20 centimeters. As the descending stream hits the water beneath, it can provoke instant ice formation.[w5] The initiating event may once again be mechanical shear, which frees protons to enter the EZ lattice to form ice.

Heat-based interpretations get even messier if you examine their dynamics. According to conventional views, latent heat is an expression of the physical transition from water to ice. In that view, a single burst of heat should evolve at the moment the water transitions to crystalline ice. However, that temporal correspondence is not seen. Instead, we found substantial delays between the appearance of latent heat and the physical transition to ice.

In one such experiment, a water droplet was placed upon a cooled metal plate. During the course of freezing, the droplet generated a flash of infrared, like the one shown in **Figure 17.10**. The flash persisted for roughly half a second before freezing began, i.e., before we could see any sign of the volume expansion that signals the physical formation of ice. We saw even longer delays with a tubular column of water standing vertically on a freezing plate. Infrared emission began at the bottom. The emission progressed upward, finally reaching the top of the column — but no volume increase was detectable at the top until approximately 1.5 seconds later. Such delays contradict the conventional expectation that latent heat and ice formation occur concurrently.

The proposed model, on the other hand, envisions ice formation as a two-stage process: first, protons flood the EZ, generating the infrared flash; second, those protons properly insert themselves between EZ planes, shifting the planes and pushing them apart. That creates ice. The triggering event *precedes* the structural event — just as the infrared evidence above implies.

This resolves the second of the two energetic issues, latent heat. I believe the proposed model conforms to all major energetic expectations.

Room Temperature Ice?

While the freezing temperature may vary under "nonstandard" conditions, to my knowledge nobody has ever seen a block of ice forming at room temperature. On the other hand, if proton invasion holds the key to ice formation, then room temperature ice falls within the realm of plausibility. Room temperature ice could form in situations where protons and EZ were in unusually ample supply and were suitably juxtaposed.

One such situation may be the water bridge (see Chapter 1). The bridge's most obvious attribute is its stiffness: hardly any droop occurs despite the bridge's several centimeter span (**Fig. 17.16**). It seems like you could almost walk across the bridge. One wonders whether the presence of an ice-like phase could account for the evident stiffness.

The bridge cross-section comprises two zones: annular and core. The core zone contains protonated water flowing from the positively charged beaker towards the negatively charged beaker. A strong infrared signal is expected from this charge flow, and that has been observed.[7] The annular zone moves in the opposite direction and more slowly. That annulus has several EZ-like features: it excludes particles; it bears negative charge (because it originates at the negative electrode); and it is birefringent,[7] which implies order.

Fig. 17.16 *Water bridge between two beakers. Bridge has a near-cylindrical cross section that comprises an annulus and a core, which cannot be distinguished in this optical image. Bridge length approximately 3 cm.*

The annular EZ with a copious supply of protons flowing nearby seems opportune for creating ice. Depending on which of those two entities is more mobile, the ice could form in the annulus, the core, or both. The ice would exist transiently; nevertheless, even transient ice, distributed throughout the bridge and fluctuating dynamically, could maintain bridge stiffness. Ice-like bridge features have been suggested before.[8]

Room temperature ice can form in configurations other than the water bridge. Ice builds, for example, when electric fields are applied across narrow gaps containing water.[9] The applied field presumably

---

Why is Water's Density Highest at 4 °C?

As water cools, its density changes. Scientists still have not explained why water's density should be highest at 4 °C. What's magical about that number?

Cooled water may contain several phases, each with a different density (*panel a*). Since EZ water is denser than bulk water (Chapters 3, 4), and bulk water is denser than ice (which floats on water), the relative amount of each phase matters. To compute overall density, you need to know how much of each phase resides in the container.

Suppose a container of water is gradually cooled. Infrared gradients increase IR throughput, which builds EZ content (**Fig. 17.4**); with a continually growing EZ fraction, the volume progressively shrinks (*panel b*). You might say the overall density has increased.

When cooling reaches a threshold, those EZs may begin transforming into ice. The transition may initially occur in localized regions, where proton concentration exceeds a threshold; patches of ice

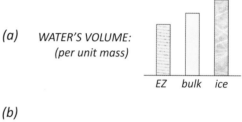

(a)   WATER'S VOLUME:
(per unit mass)

EZ   bulk   ice

(b)

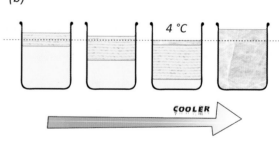

4 °C

COOLER

may begin replacing EZs. Since ice is considerably bulkier than EZ water, the overall volume may begin increasing. The temperature at which this begins can be estimated. If the massive freeze occurs near 0 °C under standard conditions, then patchy ice might begin forming a few degrees higher — plausibly 2 to 3 °C. Thus, 4 °C would be the temperature of minimum volume (third image, *panel b*). There, the density would be highest.

creates enough positive charges to convert interfacial EZs into ice, even at room temperature. All that's needed in any of these situations are ample amounts of EZ and plenty of nearby protons.

Those protons may explain other phenomena — including Mr. Mpemba's observed anomaly. Warm water contains abundant amounts of the two ingredients needed for freezing: EZ-shelled vesicles and their associated protons (Chapter 14). With those ingredients at hand, freezing the powdered ice cream and warm water mixture should not take long. Kudos to Mr. Mpemba!

Summary

The transition from water to ice requires an EZ intermediate (**Fig. 17.17**). As the water cools, EZs build (*panel i*); meanwhile, hydronium ions accumulate just beyond (*ii*). When the hydronium ion concentration reaches a critical level, protons break free and invade the negative EZ (*iii*). Those protons link adjacent EZ planes, initiating the structural transition to ice. As the process continues, the ice grows (*iv*).

This model of proton invasion resolves an energetic paradox. Creating crystalline order to form the EZ requires energy *input*. Creating crystalline order to form ice generally requires cooling, which implies energy *withdrawal*. The proton-invasion mechanism resolves the paradox: the rush of protons into the EZ delivers the potential energy of charge separation, energy that had previously been stored. In both situations, then, creating order requires energy. The energetic features of water crystallization remain consistent.

With this chapter, we conclude the scientific aspects of our inquiry into water and its phases. We've explored many of water's diverse aspects, ranging from boiling to freezing, emphasizing the central role of water's fourth phase in practically all of its behaviors.

In the chapter that follows, we end our journey by returning to the domain in which we started: the philosophical. We reflect on where we have come, what we have learned, and where we might go from there. From this vantage point, the future could hold promise for an exciting era of bold scientific progress.

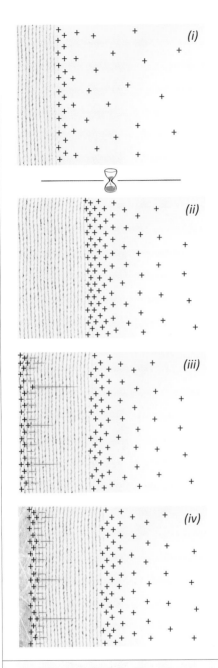

Fig. 17.17 *The ice formation mechanism. Cooling occurs from the left. IR passes from right to left through the EZ. This passage widens the EZ and separates increasing amounts of positive and negative charge. Charge invasion creates ice.*

SECTION V

Summing Up:
Unlocking Earthly Mysteries

18 The Secret Rules of Nature

While chatting with students in the laboratory several years ago, suddenly the lights went out — I almost fainted. My health had been perfectly robust — so much so that my family doctor hardly knew my name or even recognized my face. Suspecting a tumor, he now recommended a brain scan. I wound up inserted into that long scary tunnel, waiting to learn whether my life was soon to be snuffed out.

The MRI (magnetic resonance imaging) technicians showed no hint of alarm. In fact, their nonchalance left me half expecting to hear some quip about the quality of my brain. No such quip came. I did nevertheless find myself musing over what we'd recently heard about someone else's brain — a prominent US politician of rather dubious intellect: following the MRI scan of his brain, his physician allegedly reported: "Sorry sir, but it seems there's nothing right on the left side, and nothing left on the right side."

Well, some healthy grey matter apparently remained in my own brain, after all. Everything seemed normal (as far as they could tell from the MRI).

The subject of MRIs is pertinent to all you've read. The MRI machine provides a detailed image of the brain's nooks and crannies based on the relaxation properties of protons. Since the overwhelming majority of the body's protons come from water, this means the MRI measures the properties of the body's water. If water were unaffected by local structures, then the machine would produce no image; everything would look the same. The MRI can successfully visualize your brain — for better or worse — because the brain's local environment profoundly affects nearby water.

Which brings us back to the central message of this book: water participates in virtually everything. Its behavior depends on location and microenvironment, and the efficacy of MRI technology testifies to that dependence, since it relies on water's capacity to organize itself differently next to different surfaces.

After wading through 17 watery chapters, you are surely entitled to a summary of the foregoing material and an indication where it might lead. Let me begin by describing how our approach fits into the general framework of science and then proceed to substantive matters — the messages you may wish to take home from this book.

The Culture of Science

Until the modern era, scientists focused on seeking foundational mechanisms. They tried to understand how the world works. If their efforts uncovered paradigms that could explain diverse phenomena in simpler ways, then they knew they were onto something meaningful. Thus, Mendeleev's periodic table could predictably account for the multitude of known chemical reactions, and Galileo's sun-centered solar system obviated the need to invoke complex epicycles to describe planetary orbits.

The pursuit of simplicity seems to have largely evaporated from the scientific scene. In four decades of doing science, I have seen this noble culture yield to one less audacious and more pragmatic. The chutzpah has vanished. Scientists content themselves with short-term gains in narrowly focused areas rather than seeking fundamental truths that may explain broad areas of nature. A quest for detail seems to have supplanted the quest for simple unifying truths (**Fig. 18.1**).

Fig. 18.1 *Science today focuses mainly on the twigs of the tree of knowledge, attempting to add incremental detail. It assumes that supporting limbs are robust.*

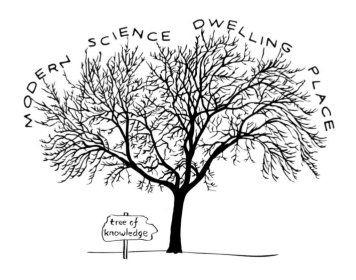

This minutae-oriented approach seems to me to bespeak a culture gone awry. You can judge this for yourself by considering the results — the scant number of conceptual revolutions that have emerged in the past three decades. I don't mean technical advances, like computers or the Internet, and I don't mean hype or *promised* revolutions, like cancer cures or endless free energy. I mean *realized conceptual revolutions* that have already *succeeded* in changing the world. How many can you identify?

Once bold, the scientific culture has become increasingly timid. It seeks incremental advances. Rarely does it question the foundational concepts on which those incremental advances are based, especially those foundational concepts that show signs of having outlived their usefulness. The culture has become obedient. It bows to the regality of prevailing dogma. In so doing, it has produced mounds of data but precious little that fundamentally advances our understanding.

I have tried to reverse this trend in these chapters by returning to the traditional way of doing science. By observing common, everyday phenomena and applying some simple logic, I have sought to answer the "how" and "why" questions that can lead to fundamental truths, while avoiding the "how much" and "what kind" questions that characterize the incremental approaches. I know it is not the fashion, but I think it offers a better path for achieving scientific progress.

The specifics of this book emerged out of a sense that something was dreadfully wrong with current thinking about water. I felt that nature should be simple at its core, yet everything I read seemed complicated. I could spout off textbook basics to anyone interested, but scratching beneath that veneer of understanding consistently exposed a substrate of questions that I found difficult to answer. That troubled me.

My search for understanding necessitated venturing into fields entirely new to me. At times, I found this unnerving, for vast bodies of knowledge seemed to lie beyond my scope of vision. On the other hand, I had the advantage of significant intellectual liberty: I wandered freely through those fields unencumbered by the constraints of the fields' orthodoxies. Few areas seemed sacred enough to remain unchallengeable.

My one goal has been to develop simple foundational principles that can lead to broad understanding. I did not pull those principles from a hat. Extracting them from the mass of relevant observations involved a long, hard journey. In the end, I believe those foundational concepts can be distilled into four central principles that govern our understanding of water.

Four Foundational Principles

Principle 1: Water Has Four Phases

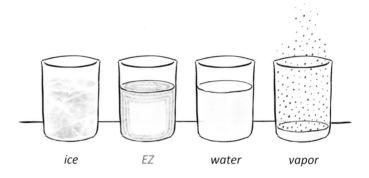

Fig. 18.2 *Water's four phases.*

From childhood, we have learned that water has three phases: solid, liquid, and vapor. Here, we have identified what might qualify as a fourth phase: the exclusion zone (**Fig. 18.2**). Neither liquid nor solid, the EZ is perhaps best described as a liquid crystal with physical properties analogous to those of raw egg white.

The term "exclusion zone" may be an unfortunate one. My friend John Watterson coined the term early on, when the most obvious feature of that zone was its exclusionary character. That definition stuck. We had fun quipping that "EZ" sounded like "easy," the opposite of hard. Hard water is full of minerals, which EZ water excludes. So the name seemed apt. In retrospect, the "liquid crystalline" phase, or the "semi-liquid" phase might have made better sense, as those descriptors fit more naturally within the phase-oriented taxonomy.

Be that as it may, the sequence of phases usually spouted off reflexively differs from what we have learned here. If the foregoing chapters offer a valid explanation of water's character, then a more appropriate phase sequence would be solid, liquid-crystalline, liquid, and vapor — *four phases, not three.*

With this fresh understanding, who knows? Undergraduates could one day find freshman chemistry far less daunting.

Principle 2: Water Stores Energy

Fig. 18.3 *The water battery.*

Water's fourth phase stores energy in two modes: order and charge separation. Order constitutes configurational potential energy, deliverable as the order gives way to disorder. For the working cell, this order-to-disorder transition constitutes a central energy delivery mechanism.[1] Charge separation, the second mode, entails electrons carrying the EZ's usual negative charge, while hydronium ions bear the corresponding positive charge. Those separated charges resemble a battery — a local repository of potential energy (**Fig. 18.3**).

Nature rarely discards repositories of available energy. It wisely parses out that energy for its diverse needs. Examples have been described throughout this book, and many more exist.

Albert Szent-Györgyi, the father of modern biochemistry, famously opined that the work of biology could be understood as the exploitation of electron energy. The EZ offers a ready source of electrons that could drive any of numerous biological reactions. The complementary hydronium ions may play an equally vital role. Positive ion concentrations build pressure, which can drive flows. Flows exist practically everywhere: in primitive and developed cells; in our circulatory systems; and in the vessels of short plants and tall trees. Hydronium ions could drive many of those flows.

The EZ's potential energy can also drive practical devices. One such device is a water purifier. Because the EZ excludes solutes, including contaminants, harvesting the EZ amounts to collecting untainted water. A simple and remarkably effective prototype has already been demonstrated.[2] It amounts to a filterless filter that achieves purification courtesy of incident electromagnetic energy.

So the potential energy associated with water's fourth phase can be exploited in different ways. Energy and water are practically synonymous. That's the reason for proposing (Chapter 7) the equation $E = H_2O$. That equation may suffer a mismatch of units, but it does capture the essence of the second principle: *water stores energy*.

Principle 3: Water Gets Energy from Light

Fig. 18.4 *The main source of electromagnetic energy on earth.*

Everyone understands that the sun illumines the earth and drives many earthly processes. What's new here is that the sun (along with, perhaps, other cosmic and earthly sources) may drive processes beyond the obvious — especially those involving water (**Fig. 18.4**).

The sun's electromagnetic energy builds potential energy in water. Photons recharge the EZ by building order and separating charge. They do this by splitting water molecules, ordering the EZ, and thereby setting up one charge polarity in the ordered zone and the opposite polarity in the bulk water zone beyond.

We don't ordinarily think of water as receiving energy. A glass of water is considered more or less in equilibrium with its environment. However, the evidence outlined in these chapters shows distinctly otherwise: a glass of water is generally far out of equilibrium. This concept may sound outlandish, but the foregoing chapters have amply demonstrated that water continually absorbs energy from the environment and transduces that energy into work.

The transduction concept may seem less exotic once you realize that plants do the same. Plants absorb radiant energy from the environment and use it for doing work. Plants, of course, comprise mostly water; therefore, it should hardly surprise that the glass of water sitting beside your potted plant may transduce incident photonic energy much like the plant does.

It may be worthwhile to take a fresh look at any scenario in which radiant energy falls incident on water. Our focus has been mainly on chemistry, but physics — and especially biology — should be considered as well. For example, when the sun breaks through the clouds, we may feel a surge of energy. That sensation surely involves our psyches; however, we may feel energized also because the incident solar energy builds real chemical energy in our cells. Some wavelengths penetrate deeply into our bodies — just place a flashlight behind the palm of your hand and watch the light penetrate all the way through to the other side.

To suggest that incident solar energy may build energy in our bodies may seem a stretch, but cells do grow faster with warmth, i.e., when exposed to infrared energy (light). Since light builds energy in water, and we are mostly water, it seems plausible that we might harvest energy from the environment. Multiple light-harvesting mechanisms can be envisioned throughout biology.

Similar principles may apply in physics and engineering. For example, harvesting light energy absorbed in water may enable the production of useful electrical energy. EZ charge separation closely resembles the initial step of photosynthesis, which entails the splitting of water next to some hydrophilic surface. This resemblance may be auspicious: if that first step works as effectively as it does in photosynthesis, then some kind of water-based harvesting of light energy may have a promising future. Designs built around water might one day replace current photovoltaic designs.

At any rate, *electromagnetic energy builds potential energy in water*, which then becomes an energy repository. That energy can radiate back toward the source from which it came, and/or it can be harvested for doing work. The energy is a gift from the environment; it is genuinely *free* energy, which we can perhaps exploit for resolving today's energy crisis.

Principle 4: Like-Charged Entities Can Attract One Another

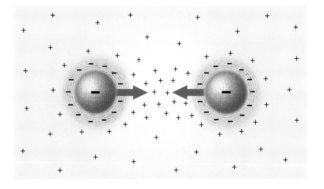

Fig. 18.5 *Mediated attraction of likes by unlikes.*

Perhaps the least obvious principle is the like-likes-like attraction (**Fig. 18.5**). The idea that like charges can attract one another seems counterintuitive until you recognize that it requires no violation of physical principles. The like charges themselves don't attract; the attraction is mediated by the unlike charges that gather in between. Those unlikes draw the like charges toward one another, until like-like repulsion balances the attraction.

Many physicists presume that like-to-like attraction cannot exist in spite of acceptance by some well-known physicists, including

Richard Feynman. Feynman coined the phrase "like-likes-like through an intermediate of unlikes." He understood that such attraction might be fundamental to physics and chemistry. Nevertheless, the majority of scientists reflexively presume that like charges must always repel. Hardly a fleeting thought is accorded the prospect that those like charges might actually *attract* if unlike charges lie in between.

This resistance may originate from the semantics: who could imagine that "like charges attract"? Surely any such phenomenon must seem like the work of the devil or, at best, of some naive charlatan. The reflexive presumption that like charges must always repel has almost certainly led to unnecessarily complex interpretations or just plain wrong answers. What could be more fundamental than the force between two charges?

This book gives substance to the early understanding of like-likes-like. It goes on to identify a source of unlike charges. Abundant unlikes come from EZ buildup, providing the ample supply of protons needed to explain the attraction.

Beyond laboratory demonstrations, the like-to-like attraction may apply broadly throughout nature, from the microscopic to the macroscopic. One possible example is in life's origin. The origin of life likely involves the concentrating of dispersed substances into condensed entities; without such condensation, no cell or pre-cell could form. The like-like-likes attraction provides a natural mechanism for mediating this kind of self-assembly: just add light, wait a bit, and voila!

Another example can be found in atmospheric clouds. Clouds are built of charged aerosol droplets. By conventional thinking, such droplets should repel and disperse; however, the like-like-likes mechanism explains why those droplets can actually coalesce into the entities that we recognize as clouds. The sun provides the energy, and the opposite charges provide the force.

Whenever like-like repulsion is proffered to explain some phenomenon, ask yourself whether the opposite — a like-to-like *attraction* — offers a better explanation. In some instances, you might find yourself walking along a fruitful path, increasing the prospect of developing a simpler and more accurate understanding of nature.

• • •

The four principles just outlined can be viewed as rules of nature, formerly obscured in some remote corner and now unveiled in a clearer light (**Fig. 18.6**).

Fig. 18.6 *Bringing hidden principles to light.*

These principles seem rich with explanatory power. They help answer simple "why" and "how" questions: Why do gels hold water? How can champagne bubbles proliferate in streams seemingly without end? How can simple hydrated wedges split apart massive boulders? How does water rise to the tops of giant redwood trees? Why do you see clouds of vapor above your hot coffee? Why does ice make you slip and fall on your face? The principles can explain many other questions whose answers have remained elusive.

Because of their vast explanatory power, I believe these four principles may prove foundational for much of nature.

Why Have These Principles Remained Secret?

If these principles are as useful as claimed, then why have they remained secret for so long? How have they escaped inclusion in the repository of common understanding?

At least four reasons come to mind.

• First, *water science has had a checkered history*. The polywater debacle left scars; it kept curious scientists away from water for decades. Any researcher confident enough to enter the arena and fortunate enough to discover something unexpected was inevitably attacked with the recycled darts used to ridicule polywater. Surely their water must have been contaminated (even though natural water is anything but pure); therefore, their results can be safely dismissed with a wave of the hand. Then came water memory. Memory stored in water seemed so improbable that it became the butt of scientific jokes: Having trouble remembering names? Try drinking more water — it will restore lost information.

Thus, the field of water was twice stung. With critics and their scorn awaiting at every turn, what prudent scientist would venture into the field of water research? Water became treacherous to study. Immersing oneself in water science has become as perilous as immersing oneself in corrosive acid.

• A second reason for the slow emergence of understanding is water's ubiquity. Water is everywhere. Water occupies a place central to so many natural processes that *few people can conceive that the basics could remain open to question.* Surely someone must have worked out those basics, probably a century or two ago. This perception keeps scientists away. If anything, their reluctance has only intensified: today's science rewards those who focus narrowly on trendy areas, leaving little room for questioning widely taught foundational science. Especially for something as deeply rooted and common as water, the incentive to question fundamentals has all but vanished.

• A third reason for the slow emergence of such fundamental principles plagues all of science: intellectual timidity. *Relying on received wisdom feels safer than dealing with the uncertainties of revolutionary disruption.* You'd think that scientists would embrace dramatic advances in fundamental science, but most of them feel more comfortable restricting themselves to minor deviations from the *status quo*. Scientists can resist revolution in the same way as any other defender of orthodoxy.

• A fourth reason is *outright fear*. Challenging received wisdom means stepping on the toes of scientists who have built careers on that wisdom. Unpleasant responses can be anticipated. For example, I have here trampled on a lot of sacred ground. I anticipate due reprimand, particularly from those scientists whose recognition, grants, patents, and other attributes of power depend on defending their scientific standing. A child might be forgiven for such apostasy; senior scientists, alas, are rarely accorded the courtesy. Thus, many career-oriented researchers maintain conservative postures, keeping their distance from anything that even smells like revolutionary challenge. That posture helps keep bread on their scientific tables.

To summarize, at least four factors bear responsibility for the painfully slow emergence of new principles: (*i*) the blighted history of the water field has kept scientists away; (*ii*) water is so common that everyone presumes that the fundamentals have been resolved; (*iii*) deviating from mainstream views can be unsettling; and (*iv*) questioning the prevailing wisdom has always been a risky business, in science as elsewhere.

These obstacles have combined to produce a long-term stall. I am trying my best to crank up that stalled engine.

The Future

We began by asking a simple question: why do exclusion zones exclude? The more we looked, the more we found. Finally, there emerged four general principles, and various insights, which you have encountered scattered throughout the book.

Seeing how far those principles can take us is a temptation to which I have admittedly succumbed. I originally intended to include material on physics and biology in this book, but readers of preliminary drafts prevailed on me to stick to water's chemistry. However, the principles elaborated here extend naturally into other scientific domains; therefore, I plan to follow up with additional books. There is much to say, particularly about physics and biology.

The key to making progress in all of these arenas must include a fresh willingness to admit that the emperor has no clothes. Even the greatest of scientific heroes might have erred. Those scientists were human: they ate the same kinds of food we eat, enjoyed the same passions we enjoy, and suffered the same frailties to which we are prone. Their ideas are not necessarily infallible. It might seem irreverent, but if we hope to penetrate toward ground truth, we need the courage to question any and all foundational assumptions, especially those that seem vulnerable. Otherwise, we risk condemning ourselves to perpetual ignorance.

Where such explorations might lead, nobody can say. Within the domain of uncertainty lies the charm of the scientific pursuit: through unfettered experimentation, logical thinking, and the occasional good luck of stumbling upon the unexpected, we may begin to illumine the dark recesses of nature.

References

Chapter 1

1. Osada Y, and Gong J (1993): Stimuli-responsive polymer gels and their application to chemomechanical systems. *Prog. Polym. Sci.*, 18, 187–226.

2. Ovchinnikova K, and Pollack GH (2009): Cylindrical phase separation in colloidal suspensions. *Phys. Rev. E.* 79(3), 036117.

3. Klyuzhin IS, Ienna F, Roeder B, Wexler A and Pollack GH (2010): Persisting water droplets on water surfaces *J. Phys. Chem.* B 114, 14020–14027.

w1. http://www.youtube.com/watch?v=FhBn1ozht-E

w2. http://www.youtube.com/watch?v=yDun7ILKrUI

w3. http://www.youtube.com/watch?v=oY1eyLEo8_A&feature=related

Chapter 2

1. Ball, Philip (1999): *H_2O: A Biography of Water.* Weidenfeld & Nicholson.

2. Ball P. (2008): Water as an Active Constituent in Cell Biology *Chem. Rev.* 108, 74–108.

3. Roy R, Tiller WA, Bell I, Hoover MR (2005): The Structure of Liquid Water: Novel Insights From Materials Research; Potential Relevance to Homeopathy. *Materials Research Innovations Online* 577-608.

4. Schiff, Michel (1995): *The Memory of Water*, Thorsens.

5. Walach H, Jonas WB, Ives J, Van Wijk R, Weingartner O, (2005): Research on Homeopathy: State of the Art. *J. Alt. and Comp Med.* 11(5) 813-829.

6. Montagnier L, Aissa J, Del Giudice E, Lavallee C, Tedeschi A and Vitiello G (2011): DNA waves and water. *J. Phys: Conf. Series* vol. 306 (on line).

w1. http://www1.lsbu.ac.uk/water/

Chapter 3

1. Henniker, JC (1949): The depth of the surface zone of a liquid. *Rev. Mod. Phys.* 21(2): 322–341.

2. Pollack, GH (2001): *Cells, Gels and the Engines of Life: A New Unifying Approach to Cell Function.* Ebner and Sons, Seattle.

3. Zheng, J -M, and Pollack GH (2003): Long-range forces extending from polymer-gel surfaces. *Phys. Rev E.* 68: 031408.

4. Zheng J -M, Chin W -C, Khijniak E, Khijniak E, Jr., Pollack GH (2006): Surfaces and Interfacial Water: Evidence that hydrophilic surfaces have long-range impact. *Adv. Colloid Interface Sci.* 127: 19–27.

5. Chai B, Mahtani A, Pollack GH (2012): Unexpected presence of solute-free zones at metal-water interfaces. *Contemporary Materials III-1-12.*

6. Zheng, J.-M., Wexler, A, Pollack, GH (2009): Effect of buffers on aqueous solute-exclusion zones around ion-exchange resins. *J. Colloid Interface Sci.* 332: 511–514.

7. Klyuzhin I, Symonds A, Magula J and Pollack, GH (2008): A new method of water purification based on the particle-exclusion phenomenon. *Environ. Sci and Techn,* 42(16) 6160–6166.

8. Yoo H, Baker DR, Pirie CM, Hovakeemian B, and Pollack GH (2011): Characteristics of water adjacent to hydrophilic interfaces. In: Water: The Forgotten Biological Molecule Ed. Denis Le Bihan and Hidenao Fukuyama, Pan Stanford Publishing Pte. Ltd. www.panstanford.com, pp. 123–136.

9. Green K, Otori T (1970): Direct measurement of membrane unstirred layers. *J Physiol (London)* 207: 93–102.

10. Ling GN (2001): *Life at the Cell and Below-Cell Level: The Hidden History of a Fundamental Revolution in Biology.* Pacific Press, NY.

11. Olodovskii PP and Berestova IL (1992). On Changes in the Structure of Water due to its contact with a solid phase. I NMR Spectroscopy Studies. *J. Eng'ng. Phys and Thermophys.* 62 622–627.

12. Yoo H. Paranji R and Pollack GH (2011): Impact of hydrophilic surfaces on interfacial water dynamics probed with NMR spectroscopy. *J. Phys. Chem Letters* 2: 532–536.

13. Bunkin NF (2011): The behavior of refractive index for water and aqueous solutions close to the Nafion interface. http://www.watercon.org/water_2011/abstracts.html

14. Tychinsky V (2011): High Electric Susceptibility is the Signature of Structured Water in Water-Containing Objects *WATER* 3, 95–99.

15. Ho, Mae-wan (2008): *The Rainbow and the Worm: The Physics of Living Organisms.* 3rd edition. World Scientific Co.

16. Roy R, Tiller, WA, Bell I, and Hoover MR (2005): The structure of liquid water; novel insights from materials research; potential relevance to homeopathy. *Mat. Res. Innovat.* 9, 98–103.

17. Ling, GN (2003): A new theoretical foundation for the polarized-oriented multilayer theory of cell water and for inanimate systems demonstrating long-range dynamic structuring of water molecules. *Physiol. Chem Phys. & Med. NMR.* 35: 91–130.

Chapter 4

1. Franks, Felix (1981): *Polywater,* MIT Press.

2. Ling, GN, (2003): A new theoretical foundation for the polarized-oriented multilayer theory of cell water and for inanimate systems demonstrating long-range dynamic structuring of water molecules. *Physiol. Chem Phys. & Med. NMR.* 35: 91–130.

3. Roy R, Tiller WA, Bell I, and Hoover MR (2005): The structure of liquid water; novel insights from materials research; potential relevance to homeopathy. *Mat. Res. Innovat.* 9–4 93124 1066–7857.

4. Ling, GN (1992): *A Revolution in the Physiology of the Living Cell.* Krieger Publ. Co, Malabar FL.

5. Pollack GH (2001): *Cells, Gels and the Engines of Life.* Ebner and Sons, Seattle, WA.

6. Lippincott ER, Stromberg RR, Grant WH, Cessac GL (1969): Polywater *Science* 164, 1482–1487.

7. Chatzidimitriou-Dreismann CA, Abdul Redah T, Streffer RMF and Mayers J. (1997): Anomalous Deep Inelastic Neutron Scattering from Liquid H2O-D2O: Evidence of Nuclear Quantum Entanglement. *Phys Rev Lett.* 79(15): 2839–2842.

8. Henderson MA (2002): The interaction of water with solid surfaces: Fundamental aspects revisited. *Surface Science Reports* 46: 1–308.

9. McGeoch JEM and McGeoch MW (2008): Entrapment of water by subunit c of ATP synthase. *Interface (Roy. Soc.)* 5(20): 311–340.

10. Kimmel GA, Matthiessen J, Baer M, Mundy CJ, Petrik NG, Smith RS, Dohnalek Z, and Kay BD (2009): No Confinement Needed: Observation of a Metastable Hydrophobic Wetting Two-Layer Ice on Graphene. *JACS* 131, 12838–12844.

11. Ji N, Ostroverkhov V, Tian CS, Shen YR (2008): Characterization of Vibrational Resonances of Water-Vapor Interfaces by Phase-Sensitive Sum-Frequency Spectroscopy, *Phys Rev Lett* 100, 096102.

12. Michaelides A and Morgenstern K (2007): Ice nanoclusters at hydrophobic metal surfaces. *Nature Mater.* 597–601.

13. Xu, K, Cao P, and Heath JR (2010): Graphene Visualizes the First Water Adlayers on Mica at Ambient Conditions. *Science* 329: 1188–1191.

14. McGeoch JEM and McGeoch MW (2008): Entrapment of water by subunit c of ATP synthase. *J. Roy Soc Interface* 5, 311–318.

15. Chai B, Mahtani AG and Pollack GH (2012): Unexpected Presence of Solute-Free Zones at Metal-Water Interfaces. *Contemporary Materials,* 3(1) 1–12.

w1. http://www.aip.org/enews/physnews/2003/split/648-1.html

Chapter 5

1. Feynman RP, Leighton RB and Sands M (1964): *The Feynman Lectures on Physics Addison-Wesley, Vol 2, Chapter 9.*

2. O'Rourke C, Klyuzhin IS, Park JS and Pollack GH (2011): Unexpected water flow through Nafion-tube punctures. *Phys. Rev. E.* 83(5) DOI: 10.1103/PhysRevE.83.056305.

3. Pollack GH (2001): *Cells, Gels and the Engines of Life.* Ebner and Sons, Seattle.

4. Pauling L (1961): A Molecular Theory of General Anesthetics. *California Institute of Technology Contribution* 2697.

5. Guckenberger R, Heim M, Cevc G, Knapp HF, Wiegrabe W, and Hillebrand A (1994): Scanning tunneling microscopy of insulators and biological specimens based on lateral conductivity of ultrathin water films. *Science* 266 (5190) 1538–1540.

6. Klimov A and Pollack GH (2007): Visualization of charge-carrier propagation in water. *Langmuir* 23(23): 11890–11895.

7. Ovchinnikova K and Pollack GH (2009): Can water store charge? *Langmuir* 25: 542–547.

Chapter 6

1. Chai B, Yoo H and Pollack GH (2009): Effect of Radiant Energy on Near-Surface Water. *J. Phys. Chem B* 113: 13953-13958.

2. Kosa T, Sukhomlinova L, Su L, Taheri B, White TJ, and Bunning TJ (2012): Light-induced liquid crystallinity. *Nature* 485 347-349.

3. Del Giudice E, Voeikov V, Teseschi A, Vitiello G (2012): Coherence in Aqueous Systems: Origin, Properties and Consequences for the Living State. Chapter 4 in "Fields of the Cell" ed. Daniel Fels and Michal Cifra, *In press.*

4. Beatty JT, Overmann J, Lince MT, Manske, AK, Lang AS, Blankenship RE, Van Dover CL, Martinson TA and Plumley FG (2005): An obligately photosynthetic bacterial anaerobe from a deep-sea hydrothermal vent. *PNAS* 102(26): 9306 – 9310.

Chapter 7

1. Pollack GH (2001): *Cells, Gels and the Engines of Life.* Ebner and Sons, Seattle.

2. Piccardi G (1962): *Chemical Basis Of Medical Climatology* Charles Thomas Publisher, Springfield, IL. *In PDF format:* http://www.rexresearch.com/piccardi/piccardi.htm

3. Rao ML, Sedlmayr SR, Roy R, and Kanzius J (2010): Polarized microwave and RF radiation effects on the structure and stability of liquid water. *Current Sci* 98 (11): 1–6.

4. Yu A, Carlson P, and Pollack, GH (2013): Unexpected axial flow through hydrophilic tubes: Implications for energetics of water. In *Water as the Fabric of Life* eds. Philip Ball and Eshel Ben Jacob. *Eur. Phys. Journal, in press.*

5. Rohani M and Pollack GH (2013): Flow through horizontal tubes submerged in water in the absence of a pressure gradient: mechanistic considerations. *Submitted for publication*

6. O'Rourke C, Klyuzhin I, Park, J-S, and Pollack GH (2011): Unexpected water flow through Nafion-tube punctures. *Phys. Rev. E.* 83(5) DOI:10.1103/PhysRevE.83.0563057. Zhao Q, Coult J, and Pollack GH (2010): Long-range attraction in aqueous colloidal suspensions. *Proc SPIE* 7376: 73716C1-C13.

w1. http://www.youtube.com/watch?v=4OkIIm5a1Lc

w2. http://www.youtube.com/watch?v=JEjFJsocDW8

Chapter 8

1. Langmuir I (1938): The Role of Attractive and Repulsive Forces in the Formation of Tactoids, Thixotropic Gels, Protein Crystals and Coacervates. *J. Chem Phys.* 6, 873–896. doi:10.1063/1.1750183

2. Feynman, RP, Leighton RB, Sands M (1964): *The Feynman Lectures on Physics*; Addison-Wesley: Reading, MA; Chapter 2, p 2.

3. Dosho S, Ise N, Ito K, Iwai S, Kitano H, Matsuoka H Okumura H and Oneo T (1993). Recent Study of Polymer Latex Dispersions. *Langmuir* 9(2): 394–411.

4. Ito K, Yoshida H, and Ise N (1994): Void Structure in Colloidal Dispersions *Science* 263 (5413) 66–68.

5. Ise N (1986): Ordering of Ionic Solutes in Dilute Solutions through Attraction of Similarly Charged Solutes — A Change of Paradigm in Colloid and Polymer Chemistry. *Angew. Chem.* 25, 323–334.

6. Ise N and Sogami IS (2005): *Structure Formation in Solutions, Ionic Polymers and Colloidal Particles,* Springer.

7. Ise N (2010): Like likes like: counterion-mediated attraction in macroionic and colloidal interaction. *Phys Chem Chem Phys,* 12, 10279–10287.

8. Mudler WH (2010): On the Theory of Electrostatic Interactions in Suspensions of Charged Colloids. *SSSAJ*: 74 (1) 1–4.

9. Ise N (2007): When, why, and how does like like like?—Electrostatic attraction between similarly charged species. *Proc. Jpn. Acad., Ser. B* (83) 192 – 198.

10. Ise N and Sogami IS (2010): Comment on "On the Theory of Electrostatic Interactions in Suspensions of Charged Colloids" by Willem H. Mulder. *SSSAJ*: 74 (1) 1–2.

11. Nagornyak E, Yoo H and Pollack GH (2009): Mechanism of attraction between like-charged particles in aqueous solution. *Soft Matter,* 5, 3850–3857.

12. Zhao Q, Zheng JM, Chai B, and Pollack GH (2008): Unexpected effect of light on colloid crystal Spacing. *Langmuir,* 24: 1750–1755.

13. Chai, B, Pollack GH (2010): Solute-free Interfacial Zones in Polar Liquids. *J Phys. Chem* B 114: 5371–5375.

14. Tata BVR, Rajamani, PV, Chakrabarti J, Nikolov A, and Wasan DT (2000): Gas-Liquid Transition in a Two-Dimensional System of Millimeter-Sized Like-Charged Metal Balls. *Phys. Rev. Letters* 84(16): 3626–3629.

15. Thornhill W and Talbott D (2007): *The Electric Universe,* Mikamar, Portland OR.

Chapter 9

1. Brush SG (1968): *A History of Random Processes.* I. Brownian Movement from Brown to Perrin. Springer, Berlin/Heidelberg.

2. Okubo T (1989a): Brownian Movement of Deionized Colloidal Spheres in Gaslike Suspensions and the Importance of the Debye Screening Length. *J Phys Chem* 93: 4352–4354.

3. Weeks ER, Crocker JC Levitt AC, Schofield A and Weitz DA (2000): Imaging of structural relaxation near the colloidal glass transition. *Science* 287: 627–31.

4. Ise N, Matsuoka, H, Ito K and Yoshida H (1990): Inhomogeneity of SoluteDistribution in Ionic Systems. *Faraday Discuss. Chem. Soc.* 90, 153–162.

5. Okubo T (1989b): Microscopic observation of gas-like, liquid-like, and crystal-like distributions of deionized colloidal spheres and the importance of the Debye-screening length. *J Chem Phys.* 90(4): 2408–2415.

6. Weeks ER and Weitz DA (2002): Properties of Cage Rearrangements Observed near the Colloidal Glass Transition. *Phys Rev Lett* 89(9): 095704.

7. Bursac P, Lenormand G, Fabry B, Oliver M, Weitz DA, Viasnoff V, Butler JP and Fredberg JJ (2005): Cytoskeletal remodelling and slow dynamics in the living cell. *Nature Materials* 4: 557–561.

8. Bhalerao A and Pollack GH (2001): Light-induced effects on Brownian displacements. *J Biophotnics* 4(3) 172–177.

9. Das R and Pollack GH (2013): Charge-based forces at the Nafion-water interface. *Langmuir, DOI: 10.1021/la304418p.*

10. Albrecht-Buehler G (2005): A long-range attraction between aggregating 3T3 cells mediated by near-infrared light scattering. *Proc Natl Acad Sci U S A.* 102(14):5050–5055.

11. Ovchinnikova, K, Pollack GH (2009): Cylindrical phase separation in colloidal suspensions. *Phys Rev E* 79(3): 036117.

12. Dosho S, Ise N, Ito K, Iwai S, Kitano H, Matsuoka H Okumura H and Oneo T (1993): Recent Study of Polymer Latex Dispersions. *Langmuir* 9(2): 394–411.

13. Chai B and Pollack GH (2010):Solute-Free Interfacial Zones in Polar Liquids. *J. Phys. Chem. B* 114: 5371–5375.

14. Zheng J-M and Pollack GH (2003): Long range forces extending from polymer surfaces. *Phys Rev E.*: 68: 031408.

Chapter 10

1. Ivanitskii GR, Deev AA and Khizhnyak EP (2005): Water surface structures observed using infrared imaging. *Physics – Uspekhi* 48(11) 1151–1159.

2. Chang H (2011): The Myth of the Boiling Point. http://www.hps.cam.ac.uk/people/chang/boiling/

3. Montagnier L, Aissa J, Ferris S, Montagnier JL and Lavall C (2009): Electromagnetic Signals are Produced by Aqueous Nanostructures Derived from Bacterial DNA Sequences. *Interdisc. Sci Comput. Life Sci* 1: 81–90. DOI 10.1007/s12539-009-0036-7.

4. Montagnier L, Aissa J, Del Giudice E, Lavallee C, Tedeschi A and Vitiello G (2011): DNA waves and water. J. Phys: Conf. Series vol. 306, 2011 (on line). http://dx.doi.org/10.1088/1742-6596/306/1/012007

5. Gurwitsch AG, Gurwitsch LD (1943): Twenty Years of Mitogenetic Radiation: Emergence, Development, and Perspectives. *Uspekhi Sovremennoi Biologii* 16:305–334. (English translation: *21st Century Science and Technology*. Fall, 1999; 12(3): 41–53).

6. Thomas Y, Kahhak L, Aissa J. (2006): The physical nature of the biological signal, a puzzling phenomenon: The critical role of Jacques Benveniste. In Pollack GH, Cameron IL, Wheatley DN, editors: *Water and the Cell*. Dordrecht: Springer, p. 325–340.

7. Chai B and Pollack GH (2010): Solute-free Interfacial Zones in Polar Liquids. *J Phys. Chem B* 114: 5371–5375.

Chapter 11

1. Lin SC, Lee WI and Schurr JM (1978): Brownian motion of highly charged poly(L-lysine). Effects of salt and polyion concentration. Biopolymers 17(4) 1041–1064.

2. Weiss M, Eisner M, Kartberg F, Nilsson T (2004): Anomalous Subdiffusion is a Measure for Cytoplasmic Crowding in Living Cells.Biophys J, 87(5): 3518–3524.

3. Halliday I (1963): Diffusion effects observed in the wake spectrum of a Geminid meteor. Smithsonian Contrib to Astrophys, 7, 161–169.

4. Chai B, Zheng JM, Zhao Q, and Pollack GH (2008): Spectroscopic studies of solutes in aqueous solution. *J Phys Chem A* 112: 2242–2247.

5. Sedlák M (2006): Large-Scale Supramolecular Structure in Solutions of Low Molar Mass Compounds and Mixtures of Liquids: I. Light Scattering Characterization, *J. Phys. Chem. B*, 110 (9), 4329–4338.

6. Zhao Q, Ovchinnikova K, Chai B., Yoo H, Magula J and Pollack GH (2009): Role of proton gradients in the mechanism of osmosis. *J Phys Chem B* 113: 10708–10714.

7. Zheng JM and Pollack GH (2003): Long range forces extending from polymer surfaces. *Phys Rev E.*: 68: 031408.

8. Loeb J (1921): The Origin of the Potential differences responsible for anomalous osmosis. *J Gen Physiol* 20; 4(2): 213–226.

9. Osada Y, and Gong J (1993): Stimuli-responsive polymer gels and their application to chemomechanical systems. *Prog. Polym. Sci.*, 18, 187–226.

w1. http://news.softpedia.com/news/Unmixed-Pool-of-Freshwater-Found-in-Arctic-Ocean-193373.shtml

w2. http://vivithemage.com/zen/blog/picturesaroundthe world/13.jpg.php

Chapter 12

1. Tada T, Kanekoa D, Gong, J -P, Kanekoa T, and Osada Y (2004): Surface friction of poly(dimethyl siloxane) gel and its transition phenomenon. *Tribology Letters*, Vol. 17, No. 3.

2. Stern KR, Bidlack J, Jansky, SH (1991): *Introductory Plant Biology*, McGraw Hill.

3. So E, Stahlberg R, and Pollack GH (2012): Exclusion zone as an intermediate between ice and water. in: *Water and Society*, ed. DW Pepper and CA Brebbia, WIT Press. pp 3 – 11.

4. Chai B, Mahtani AG, and Pollack GH (2012): Unexpected Presence of Solute-Free Zones at Metal-Water Interfaces. *Contemporary Materials*, III (1) 1–12.

5. Chai B, Mahtani A, Pollack GH (2013): Influence of Electrical Connection between Metal Electrodes on Long Range Solute-Free Zones. *Submitted for publication*.

6. Musumeci F and Pollack GH (2012): Influence of water on the work function of certain metals. *Chem Phys Lett* 536: 65–67.

7. O'Rourke C, Klyuzhin IS, Park JS, and Pollack, GH (2011): Unexpected water flow through Nafion-tube punctures. *Phys. Rev. E.* 83(5) DOI:10.1103/PhysRevE.83.056305.

w1. http://en.wikipedia.org/wiki/Dead_water
http://www.youtube.com/watch?v=PCOL8kUtufg

Chapter 13

1. Klyuzhin, IS, Ienna, F, Roeder B, Wexler, A and Pollack GH: Persisting Water Droplets on Water Surfaces (2010): *J. Phys Chem B* 114:14020–14027.

2. Melehy, M. (2010): *Introduction to Interfacial Transport.* Author House, Bloomington IN.

3. Chai B, Zheng JM, Zhao Q, and Pollack GH (2008): Spectroscopic studies of solutes in aqueous solution. *J. Phys. Chem A* 112:2242–2247.

4. Bunkin, NF, Suyazov NV; Shkirin AV, Ignatiev PS, Indukaev KV (2009): Nanoscale structure of dissolved air bubbles in water as studied by measuring the elements of the scattering matrix. *J Chem Phys* 130, 134308.

5. Zheng JM and Pollack GH (2003): Long range forces extending from polymer surfaces. *Phys Rev E.*: 68:031408.

6. Ninham, BW and Lo Nostro, P (2010): *Molecular Forces and Self Assembly in Colloid, Nano Sciences, and Biology.* Cambridge University Press.

7. Zuev AL, and Kostarev KG (2008): Certain peculiarities of solutocapillary convection. *Physics - Uspekhi* 51 (10) 1027–1045.

8. Hu W, Ishii, KS and Ohta AT (2011): Micro-assembly using optically controlled bubble microrobots. *Appl. Phys. Lett.* 99 (094103), 1–3.

9. Spiel DE (1998): On the births of film drops from bubbles bursting on seawater surfaces. *J. Geophys Res* 103 (C11) 24,907–24918.

w1. http://www.youtube.com/watch?v=lkqSEApCSvM&feature=related

Chapter 14

1. Chang H. (2011) http://www.hps.cam.ac.uk/people/chang/boiling/

Chapter 15

1. Ienna F, Yoo H and Pollack GH (2012): Spatially Resolved Evaporative Patterns from Water. *Soft Matter* 8 (47), 11850 – 11856.

2. Ivanitskii GR, Deev AA, Khizhnyak EP (2005): Water surface structures observed using infrared imaging *Physics ± Uspekhi* 48 (11) 1151–1159.

3. Chai B, Zheng JM, Zhao Q, and Pollack GH (2008): Spectroscopic studies of solutes in aqueous solution. *J. Phys. Chem., A* 112 2242–2247.

4. Tychinsky V (2011): High Electric Susceptibility is the Signature of Structured Water in Water-Containing Objects. *WATER* 3: 95–99.

5. Bunkin N (2013): Refractive index of water and aqueous solutions in optical frequency range close to Nafion interface. (*submitted*).

w1. http://www.youtube.com/watch?v=bT-fctr32pE

w2. http://touristattractionsgallery.com/niagara-is-the-largest-waterfall-in-the-world

w3. http://www.engineeringtoolbox.com/air-composition-d_212.html

Chapter 16

1. Ovchinnikova K and Pollack GH (2009): Cylindrical phase separation in colloidal suspensions. *Phys. Rev. E* 79 (3) 036117.

2. Mopper K and Lindroth P (1982): Diel and depth variations in dissolved free amino acids and ammonium in the Baltic Sea determined by shipboard HPLC analysis. *Limnol. Oceanogr.* 27(2): 336–347.

3. Pollack GH (2001): *Cells, Gels, and the Engines of Life.* Ebner and Sons, Seattle.

4. Cameron I (2010): Dye Exclusion and Other Physical Properties of Hen Egg White. *WATER* (2): 83–96.

5. Gaddis, V. (1965): *Invisible Horizons: True Mysteries of the Sea*. Chilton Books, Phila.

6. Heilbron JL (1999): Electricity in the 17th & 18th Centuries:: A Study in Early Modern Physics (Dover Books on Physics), p 239.

7. Wegner LH and Zimmermann U (2004): Bicarbonate-Induced Alkalinization of the Xylem Sap in Intact Maize Seedlings as Measured in Situ with a Novel Xylem pH Probe. *Plant Physiol*. 136(3): 3469–3477.

8. Klyuzhin IS, Ienna F, Roeder B, Wexler A, and Pollack GH (2010): Persisting water droplets on water surfaces. *J. Phys. Chem. B* 114: 14020–14027.

w1. http://www.youtube.com/watch?v=45yabrnryXk

w2. http://oai.dtic.mil/oai/oai?verb=getRecord&metadata Prefix=html&identifier=ADB228588

w3. http://www.bermuda-triangle.org/html/don_henry.html

w4. http://www.youtube.com/watch?v=yf1n00LW1XI&feature=related

w5. http://www.youtube.com/watch?v=DkIIMEVnNDg

Chapter 17

1. Mpemba EB, Osborne D G. (1969): "Cool?" *Physics Education* (Institute of Physics) : 172–175. doi:10.1088/0031-9120/4/3/312.

2. Mpemba EB and Osborne DG (1979): The Mpemba effect. Physics Education (Institute of Physics) : 410–412. doi:10.1088/0031-9120/14/7/312. http://www.iop.org/EJ/article/0031-9120/14/7/312/pev14i7p410.pdf.

3. So E, Stahlberg R, and Pollack GH (2012): Exclusion zone as an intermediate between ice and water. in: *Water and Society*, ed. DW Pepper and CA Brebbia, WIT Press, pp 3-11.

4. Hori T. (1956): *Low Temperature Science A* 15:34 (English translation) No. 62, US Army Snow, Ice and Permafrost Res. Establishment, Corps of Engineers, Wilmette, Ill

5. Ehre D, Lavert E, Lahav M, Lubomirsky I (2010): Water Freezes Differently on Positively And Negatively Charged Surfaces of Pyroelectric Materials *Science* 327: 672–675.

6. Workman, EJ and Reynolds SE (1950): Electrical Phenomena Occurring during the Freezing of Dilute Aqueous Solutions and Their Possible Relationship to Thunderstorm Electricity. *Phys. Rev*. 78(1): 254–260.

7. Woisetschlager J, Gatterer K, and Fuchs EC (2010): Experiments in a floating bridge *Exp Fluids* 48: 121–131.

8. Piatkowski L, Wexler AD, Fuchs EC, Schoenmaker H and Bakker HJ (2012): Ultrafast vibrational energy relaxation of the water bridge *Phys. Chem. Chem. Phys. 14(18): 6160-6164.*

9. Choi E-M, Yoon Y-H, Lee S, Kang H (2005): Freezing Transition of Interfacial Water at Room Temperature under Electric Fields. *Phys. Rev. Lett*. 95, 085701.

w1. http://www.youtube.com/watch?v=Gp8vc0DWf3U

w2. http://www.youtube.com/watch?v=ywh5TQ5B4Es

w3. http://www.youtube.com/watch?v=bDwZqBqrLQ&p =2556DBFD5031F40F&index=39

w4. http://www.youtube.com/watch?v=xuhUTaFmaX8&N R=1&feature=endscreen

w5. http://www.youtube.com/watch?v=fSPzMva9_CE

Chapter 18

1. Pollack GH (2001): *Cells, Gels and the Engines of Life: A New Unifying Approach to Cell Function*. Ebner and Sons, Seattle.

2. Klyuzhin I, Symonds A, Magula J and Pollack GH (2008): A new method of water purification based on the particle-exclusion phenomenon. *Environ Sci and Technol* 42(16): 6160–6166.

Photo Credits

Chapter 1

Figure 1.4: Elmar Fuchs

Chapter 2

Figure 2.3: With permission

Chapter 3

Figure 3.19: With permission, Dr. Maewan Ho

Chapter 5

Figure 5.1: Basil Hovakeemian
Figure 5.5: Li Zheng and Ronnie Das
Figure 5.6: Hyok Yoo
Box (Nerves, Pain, Anesthesia): Nenad Kundacina
Figure 5.8: With permission

Chapter 6

Figure 6.1: Binghua Chai
Figure 6.3: Bora Kim
Figure 6.5: With permission
Figure 6.12: With permission, Office of Ocean Exploration and Research, NOAA

Chapter 7

Figure 7.6: With permission

Chapter 8

Figure 8.4: With permission
Figure 8.11: With permission

Chapter 9

Figure 9.1: With permission
Figure 9.4: With permission
Figure 9.8: Derek Nhan
Figure 9.9: Ronnie Das
Figure 9.10: Rainer Stahlberg
Figure 9.12: Kate Ovchinnikova

Chapter 10

Figure 10.6: Eugene Khijniak
Figure 10.9: Anna Song
Figure 10.10: Anna Song

Chapter 11

Figure 11.2: Ronnie Das

Chapter 13

Figure 13.3: With permission
Figure 13.6: Sudeshna Sawoo
Figure 13.7: Rolf Ypma, Orion Polinsky
Figure 13.8: Georg Schröcker
Figure 13.9: Eric Gupta
Figure 13.10: Hyok Yoo

Chapter 14

Figure 14.5: George Danilov
Figure 14.6: Hyok Yoo
Figure 14.7: Hyok Yoo
Figure 14.10: George Danilov
Figure 14.11: Rainer Stahlberg

Chapter 15

Figure 15.1: Ethan Pollack
Figure 15.2: With permission
Figure 15.3: With permission
Figure 15.4: With permission
Figure 15.5: Hyok Yoo, Ethan Pollack
Figure 15.6: Federico Ienna
Figure 15.7: ZiYao Wang
Figure 15.8: Yan Dong
Figure 15.9 (a): Federico Ienna
Figure 15.9 (b): ZiYao Wang
Figure 15.10: Federico Ienna
Figure 15.11: Federico Ienna
Box (Droplet Repulsion, Kelvin Water Dropper): Zheng Li

Chapter 16

Figure 16.3: Laura Marshall, Hyok Yoo
Figure 16.14: Patrick Belenky
Figure 16.16: Ivan Klyuzhin
Figure 16.17: With permission

Chapter 17

Figure 17.3: Hyok Yoo
Figure 17.5: With permission
Figure 17.7: Rainer Stahlberg
Figure 17.8: Rainer Stahlberg
Figure 17.9: Rainer Stahlberg
Figure 17.10: Hyok Yoo
Figure 17.16: Elmar Fuchs

Glossary

Annulus: A ring-like structure.

Anode: An electrode through which electric current flows *into* a polarized electrical device. (By convention, the direction of electric current is opposite to the direction of electron flow.)

Asperity: Surface unevenness.

Bernoulli hump: Ocean surface bulges created by an object moving underwater.

Birefringence: An optical property of a material in which the refractive index depends on direction. Crystalline minerals such as calcite and quartz show birefringence.

Catalysis: The increase in rate of a chemical reaction arising from the participation of a substance called a catalyst. The catalyst is not consumed by the reaction.

Cathode: An electrode through which electric current flows *out of* a polarized electrical device. (By convention, the direction of electric current is opposite to the direction of electron flow.)

Colligative properties: Referring to the physical properties (e.g., freezing and boiling points) of solutions. Colligative properties depend on the ratio of solute particles to solvent molecules, and are largely independent of the nature of the solute.

Colloid: A substance evenly dispersed throughout another substance; the dispersed substances are often particles ranging from 1 nm to 1000 nm.

Cuvette: A small tube with square or circular cross section, sealed at one end, used for spectroscopic measurements.

Daltons: The standard units used for indicating mass on an atomic or molecular scale.

DC: Direct Current. Refers to the unidirectional flow of charge – as opposed to AC, or alternating current, where charge flow periodically reverses direction.

Dewar: A flask designed to provide good thermal insulation. When filled with a hot (or cold) liquid, the liquid will stay hot (or cold) for much longer than in a typical container.

Dielectric: An electrical insulator that can be polarized. When a dielectric is placed in an electric field, electric charges do not flow through the material as they do in a conductor, but instead slightly shift, creating more positives on one side and more negatives on the other.

Dipole: A separation of positive and negative charges. The simplest example is a pair of electric charges of equal magnitude but opposite sign, separated by some (usually small) distance.

Diurnal: Daily. A diurnal cycle refers to any pattern that recurs daily.

Electrode: An electrical conductor used to make contact with a device or material. Current enters or leaves the circuit through the electrodes.

Electromagnetic spectrum: The range of all possible frequencies of electromagnetic radiation.

Electronegative: Having a negative charge; assuming negative potential when in contact with a dissimilar substance; also, the tendency of an atom or functional group to attract electrons.

Electrostatic: Phenomena arising from the forces that electric charges exert on each other.

Enthalpy: A measure of the total energy of a thermodynamic system; the amount of heat used or released in a system at constant pressure.

Entropy: An expression of disorder or randomness; a measure of a system's thermal energy that is unavailable for doing useful work.

Faraday cage: An enclosure built from a conducting material or mesh, used to block external electric fields.

Fenestration: Openings in a structure.

Filament: In biology, a long chain of proteins.

Fluorescence: The emission of light by a substance that has absorbed light or other electromagnetic radiation.

Frictional coefficient (coefficient of friction): Between two bodies, the ratio of the force of sliding friction and the force pressing those bodies together.

Gas clathrate: Crystalline solids resembling ice, in which gas molecules are trapped inside cages of water molecules.

Gradient: The variation in space of any quantity that can be represented by a slope. The gradient represents the steepness and direction of that slope.

Heat capacity: The amount of heat required to raise the temperature of a substance by one degree Celsius.

Hexamer: Something composed of six subunits.

Homogeneous: Uniform in composition.

Hydration: The supply and retention of water.

Hydronium ion (H_3O^+): An ion produced by the protonation of water.

Hydroxyl ion (OH^-): A negatively charged ion containing an oxygen atom covalently bonded to a hydrogen atom.

Incident: In physics, something that strikes a surface, e.g., an incident ray of light.

Interfacial: Relating to a boundary between two portions of matter or space.

Ionosphere: The part of the upper atmosphere ionized by solar radiation.

Lattice: A regular, periodic configuration of particles or objects distributed throughout an area or space, especially the arrangement of ions or molecules in a crystalline solid.

Light-emitting diode (LED): A semiconductor light source.

Luminol ($C8H7N3O2$): A versatile chemical that exhibits blue luminescence when mixed with an appropriate oxidizing agent.

Mean square: The average of the squares of a set of numbers.

Nucleate: To provide a nucleus or starting point for something.

Osmosis: The movement of solvent molecules, usually through a membrane, to a region of higher solute concentration.

Oxide: A chemical compound that contains at least one oxygen atom and some other element.

Photoelectric effect: Emission of electrons from matter as a result of absorption of electromagnetic energy.

Photon: A minute energy packet of electromagnetic radiation, often used with reference to light.

Pole: In batteries, the two terminals, positive and negative.

Polyacrylic acid: Generic name for synthetic high molecular weight polymers of acrylic acid.

Polymer: A large molecule consisting of repeating subunits.

Precipitation: The formation of solid material in a solution, usually settling to the bottom.

Pyroelectric: In chemistry, the tendency of certain materials to generate charge when heated or cooled.

Raman spectroscopy: A technique used to study vibrational, rotational, and other low-frequency vibrational modes in a system.

Refractive index: A number that describes how light, or any other electromagnetic radiation, propagates through a medium.

Semiconductor: A material whose electrical conductivity lies between a conductor and an insulator.

Solvent: A substance that dissolves a solute.

Stoichiometric complex: Referring to a fixed ratio of components of a substance.

Stoichiometry: Referring to the relative quantities of reactants and products of a chemical reaction, usually given in whole numbers.

Surface energy: The excess energy at the surface of a material compared to the bulk.

Thermodynamics: The branch of natural science concerned with heat and its relation to other forms of energy and work.

Thixotropy: A characteristic of certain gels that may flow when disturbed by sufficient shaking or shearing.

Transducer: A device that converts one form of energy into another.

Triboelectric effect: A type of contact electrification in which certain materials become electrically charged after rubbing against a different material.

Vesicle: A small sac, especially one containing fluid.

Work: Originally "weight lifted through a height", but more generally the product of force and the resulting displacement in the direction of that force.

Index

Page numbers in *bold-italic* denote figures.

A

acidity
  effects on EZ 97
acoustic energy 91, 119. *See also* ultrasound
adhesion 205
  to ice 208
air
  flow resistance 271, *271*, 272, *272*
  friction 278
  gas ratios 273
    clathrates 274, *274*
  linkages 271, 272, 273, 274
    vesicles 275, *275*
  radio transmission 275, 276
anesthetics 77, *77*
Archimedes 292, 293
atmospheric coupling 276, 277, 278

B

bacteria 153, *153*
Ball, Philip 13, 14, 16, 21
batteries 211, *211*, 212
  external inputs 212, 213, 214
  EZ mechanism *214, 216*
  in cars 215
  theory 211
  water. *See* EZ charge separation
Benveniste, Jacques 19, 20, 21, 22
Bermuda Triangle 294, *294*
Bernoulli humps 290
blood flow 26, 116
blood vessels 32
boiling
  mechanism 247, *247*, 248
  sounds 249

  temperature 248
Brownian motion 115, 141
  conflicts 143, 144, 145, 147
  Einstein theory 142, *142*, 143, 144, 146, 156, 159, 160, *160*
  equation 156
  light-driven 149, *150*, 152, 154, 155, *155*, 156, 158, 159, *160*
  light effects 145, *145*, 146
  random walks 142, 143, *143*, 184, *184*
  salt solutions 144, 160
  synchronization 145, 156
  temperature dependence 143
Brown, Robert 141, *141*
bubbles xxv, 221, *221*, 222. *See also* vesicles
  ambiguity 222, 223, *223*, 231
  attraction to light 230
  bubble guillotine 223, *223*, 224
  charge 240
  clusters 225, 229
  formation 221, 225, 234
  infrared emission 240, *240, 241*
  membrane 223, 224, 228
    EZ 228, 230
  ordering 228, 229, *229*
  UV absorbance 228
bulk water xxii
buoyancy 292
  cohesion *292*, 293
  current theory 293
  sinking 294
  surface EZs 293, *293*

C

Canny, Martin 299
capillary action 294, 295
  current theory 295, *295*
  EZ mechanism 296, *296*, 297, 298

  in plants 299, 300
  meniscus 295, 298
cartilage 206
catalysis 214, *214*, 215, 216
champagne 238, *238*
Chaplin, Martin 15
charged particles 125, *126*, 127, *131, 132*, 138. *See also* like likes like
  beads 129, *129*, 130, *130, 131*
  end point 132, 133
  EZ mediation 130, 131, *131*, 132
charge oscillation 165, 168
  in EZs 168
  in water 169
clouds 267, 337
  formation 6, *6*
  in infrared 167, *167*
coffee 255
colligative properties 312
colloids 17, 126, *126*, 133
  crystals 127, *127*, 133, 136, 139, 145, 157, *157*, 188
concrete 178
condensation 233, 250
contact angles 5, 243, *243*
contaminants 18, 19, 46, 48, 49
  from gels 31
controversies
  polywater 18, 19, *19*, 45, 46, 47, 48, 49, 55
  water memory 19, 21, *21*
convection. *See also* temperature artifacts
crystals 32, 176, 321, 321
  exclusion 33
  nucleation 32
  sugar 135, *321*